Patch Clamping

Patch Clamping

An Introductory Guide to Patch Clamp Electrophysiology

Areles Molleman
University of Hertfordshire, UK

JOHN WILEY & SONS, LTD

Copyright © 2003   John Wiley & Sons Ltd, The Atrium, Southern Gate, Chichester, West Sussex
PO19 8SQ, England

Telephone (+44) 1243 779777

Email (for orders and customer service enquiries): cs-books@wiley.co.uk
Visit our Home Page on www.wileyeurope.com or www.wiley.com

Reprinted April 2005, January 2006, September 2006, May 2007, February 2008, September 2009

All Rights Reserved. No part of this publication may be reproduced, stored in a retrieval system or transmitted in any form or by any means, electronic, mechanical, photocopying, recording, scanning or otherwise, except under the terms of the Copyright, Designs and Patents Act 1988 or under the terms of a licence issued by the Copyright Licensing Agency Ltd, 90 Tottenham Court Road, London W1T 4LP, UK, without the permission in writing of the Publisher. Requests to the Publisher should be addressed to the Permissions Department, John Wiley & Sons Ltd, The Atrium, Southern Gate, Chichester, West Sussex PO19 8SQ, England, or emailed or permreq@wiley.co.uk, or faxed to (+44) 1243 770571.

This publication is designed to provide accurate and authoritative information in regard to the subject matter covered. It is sold on the understanding that the Publisher is not engaged in rendering professional services. If professional advice or other expert assistance is required, the services of a competent professional should be sought.

Other Wiley Editorial Offices

John Wiley & Sons Inc., 111 River Street, Hoboken, NJ 07030, USA

Jossey-Bass, 989 Market Street, San Francisco, CA 94103-1741, USA

Wiley-VCH Verlag GmbH, Boschstr. 12, D-69469 Weinheim, Germany

John Wiley & Sons Australia Ltd, 33 Park Road, Milton, Queensland 4064, Australia

John Wiley & Sons (Asia) Pte Ltd, 2 Clementi Loop #02-01, Jin Xing Distripark, Singapore 129809

John Wiley & Sons Canada Ltd, 22 Worcester Road, Etobicoke, Ontario, Canada M9W 1L1

Library of Congress Cataloging-in-Publication Data

Molleman, Areles.
  Patch clamping : an introductory guide to patch clamp electrophysiology / Areles Molleman.
    p. cm.
  Includes bibliographical references and index.
  ISBN 0 471 48685 X
  1. Patch-clamp techniques (Electrophysiology)   I. Title.

QP341.M684 2002
572'.437 — dc21

2002027413

British Library Cataloguing in Publication Data

A catalogue record for this book is available from the British Library

ISBN 978-0-471-48685-5 (PB)

Typeset in 11/13½ Sabon by Keytec Typesetting Ltd
Printed and bound in Great Britain by TJ International, Padstow, Cornwall

Contents

Preface	ix

1 Introduction — 1
1.1 Patch Clamping and its Context — 1

2 Basic Theoretical Principles — 5
2.1 Introduction to Membrane Biology — 5
    2.1.1 The plasma membrane and its ionic environment — 5
    2.1.2 Electrochemical gradients and the Nernst equation — 7
    2.1.3 Maintenance of ion gradients and the membrane potential — 8
    2.1.4 Ion channels — 11
2.2 Electrical Properties of the Cell Membrane — 13
    2.2.1 Driving force and membrane resistance — 13
    2.2.2 Membrane capacitance — 15
    2.2.3 Consequences of membrane capacitance — 16
    2.2.4 An electronic model of the plasma membrane — 17
2.3 Recording Modes and their Equivalent Circuits — 18
    2.3.1 The basics of equivalent circuits — 18
    2.3.2 Intracellular recording — 22
    2.3.3 Voltage clamp and current clamp — 28
    2.3.4 Introduction to patch clamp configurations — 30
    2.3.5 The equivalent circuit for the cell-attached patch configuration — 35
    2.3.6 The equivalent circuit for the whole-cell configuration — 39
    2.3.7 The equivalent circuit for the excised patch configurations — 40

3 Requirements — 43
3.1 The Platform — 43
    3.1.1 Stability: vibrations and drift — 43
    3.1.2 Where in the building should the set-up be placed? — 44
    3.1.3 Anti-vibration tables — 45
3.2 Mechanics and Optics — 47
    3.2.1 The microscope — 48

	3.2.2	Micromanipulators		52
	3.2.3	Pipette pressure		56
	3.2.4	Baths and superfusion systems		57
3.3	Electrodes and Micropipettes			64
	3.3.1	Solid–liquid junction potentials and polarisation		65
	3.3.2	The bath electrode		67
	3.3.3	Micropipettes		67
	3.3.4	Liquid junction potentials		74
3.4	Electronics			75
	3.4.1	External noise and Faraday cages		76
	3.4.2	Patch clamp amplifiers		81
	3.4.3	Noise prevention and signal conditioning		84
	3.4.4	Data acquisition and digitisation		90
	3.4.5	Computers and software		93

4 The Practice of Patch Clamping — 95

4.1	Preparing the Experiment and Making a Seal		95
	4.1.1	Setting up	95
	4.1.2	Bringing the pipette near the preparation	98
	4.1.3	Making the seal	101
4.2	Whole-cell Modes		104
	4.2.1	Conventional whole-cell recording	104
	4.2.2	Perforated patch recording	108
4.3	Single-channel Modes		110
	4.3.1	General notes	110
	4.3.2	Cell-attached patch	112
	4.3.3	Excised patches	113

5 Whole-cell Protocols and Data Analysis — 116

5.1	Standard Cellular Parameters		116
5.2	Voltage-activated Currents		116
	5.2.1	Introduction to pulse protocols	116
	5.2.2	Signal conditioning and positive/negative subtraction	119
	5.2.3	Space clamp artefacts	123
	5.2.4	Isolation of a homogeneous population of channels	126
	5.2.5	Current–voltage relationships and reversal potential	127
	5.2.6	Determination of relative permeabilities	131
	5.2.7	Activation and inactivation studies	132
5.3	Non-voltage-activated Currents		137
	5.3.1	Introduction to continuous recording	137
	5.3.2	Determination of reversal potential using voltage ramps	138

6 Single-channel Protocols and Data Analysis — 141

6.1	General Single-channel Practice and Analysis		141
	6.1.1	Practical notes	141
	6.1.2	Amplitude analysis	143
	6.1.3	Event detection	148

	6.1.4 Dwell time analysis	152
6.2	Continuous Recording of Single Channels	157
	6.2.1 Data acquisition	157
	6.2.2 Spontaneous activity	158
	6.2.3 Receptor-induced activity	160
6.3	Study of Single-voltage-dependent Channels	160
	6.3.1 Step protocols	160
	6.3.2 Ramp protocols	162
	6.3.3 Correlation with macro-currents	164

Further Reading 167

Index 171

Preface

Since the advent of patch clamp electrophysiology at the end of the 1970s, there has been an ever-increasing interest in the application of its power to investigate ion-channel-related bioscientific questions in unprecedented detail. Whereas patch clamping was pioneered in specialist biophysical laboratories, later there was an expansion to many other fields in biology, as well as to basic research in medicine and related areas. This has not occurred without problems: patch clamping was developed by biologists who were strongly orientated towards physics and who were collaborating closely with engineers and physicists. In the expansion to traditionally less-biophysical fields it is very apparent that concepts and methods in patch clamping are fundamentally different from the mostly chemistry-based work that is performed there. Many undergraduate biosciences programmes have been slow to react to this situation, sometimes owing to the lack of experienced lecturers. Even if expert teachers are available, the absence of an appropriate textbook makes it hard for the teachers to communicate the basic concepts. Consequently, it is often difficult for researchers to recruit workers with the appropriate training to perform patch clamping. At the same time researchers who are considering using patch clamping in their work can find it hard to evaluate the implications if they lack training and experience themselves.

Textbooks on patch clamping are too general for practical laboratory use or are written for experienced biophysicists. Most of them also have a format in which each chapter is written by different authors who write about their respective specialist areas, which inevitably means that the books lack continuity if not coherence. This book is intended to provide, in a single coherent volume, the basic knowledge and an overview of the materials, skills and procedures required to start patch clamping successfully. The book is written for those who are novices at patch clamping but have some background in Biology or Medical Sciences, be they final-year undergraduate or MSc/PhD students, post-doctoral researchers or established workers from other fields who may require help in the concepts of electronics, for example. A portion of the book is therefore spent on this subject with the emphasis on relevance to membrane biology. The book also should be useful as a basic method reference, particularly in relation to data acquisition protocols and analysis.

One of the developments since the expansion of patch clamping into a wider field has been that equipment and software have become much more commercially available and 'user-friendly'. The unfortunate side-effect of this has been that it has become easier to work with the equipment, obtain data and have them analysed automatically without the experimenter evaluating the validity of the results or the analysis. Although this is a problem with all automated systems, from statistics software to nuclear magnetic resonance spectral analysis, in patch clamping there are many stages in which input (judgement) from the experimenter is required for proper application of the equipment and analysis techniques. The right input comes with experience and appreciation of the experimental situation. This book is meant to provide the start of that appreciation.

The book is not a dictionary of patch clamping. The main fundamentals are touched upon, but the level of discussion is measured to convey the concepts in enough depth to apply in practice. There should then be enough understanding to provide access to specialist literature if required. A selection of this literature is listed in a 'Further Reading' section.

The success of this text should be measured by its ability to minimise the number of other books that need to be studied, wheels reinvented and time spent to master basic ideas before successful patch clamp recordings are made. My hope is that this book will allow patch clamping to become a rewarding pursuit for many more scientists.

Areles Molleman

1
Introduction

1.1 Patch Clamping and its Context

The historical route to present-day patch clamping started with the scientific recognition that electrical phenomena are part of animal physiology. This bioelectricity was demonstrated in the nineteenth century in frogs, where muscle movements could be evoked by applying electrical stimuli to the animal. The recording of inherent electrical activity can be charted by the development of increasingly sophisticated electrodes. Long before anything was known about ion channels, membrane potentials were recorded using crude glass electrodes in the few preparations where this is possible, most famously the squid giant axon (Hodgkin and Huxley, 1952). To measure membrane potential in other preparations, much finer electrodes were needed. Graham and Gerard (1946) produced glass micropipettes with tips of 2–5 μm diameter that could be used to measure resting membrane potentials in skeletal muscle cells. Unlike the squid giant axon, this involved puncturing the cell membrane. It was found that even smaller tip pipettes greatly increased the success ratio and consistency in results (Ling and Gerard, 1949). Applications for intracellular recording expanded to include the study of synaptic potentials and other changes in membrane potential, while glass micropipettes were also starting to be used for superfusion and application of drugs.

Around the same time the principle of voltage clamp was developed by Cole (1949) for the squid giant axon. The further significance of Hodgkin and Huxley's work (1952) lies in the full fruition of voltage clamp application and analysis to describe action potential conductances. However, the application of voltage clamp in cells using micropipettes for intracellular recording had to wait for another 26 years, when voltage clamp was used in two-electrode mode (Meech and Standen, 1975) or single-electrode mode (Wilson and Goldner, 1975) (see also Section 2.3.3).

Neher and Sakmann (1976) brilliantly used the strengths of microelectrodes and voltage clamp but avoided the complicated voltage clamp problems by using a relatively large-bore pipette that does not penetrate the cell but forms a very tight seal with it. The adhesion between the pipette glass and the membrane is actually stronger than the membrane, so that pulling away the pipette will break the membrane around the patch but keep the seal intact. The key paper that is usually referenced in patch clamp work summarises the key elements of the technique (Hamill *et al.*, 1981). The main configurations are already present in this paper, demonstrating that the inventors were immediately aware of the great possibilities of patch clamping. Sakmann and Neher received the Nobel Prize in Physiology and Medicine for their work in 1991. Such has been the success of patch clamping that 'conventional' intracellular recording has drifted somewhat into the background, although it still has an important role to play in electrophysiology (see Section 2.3.2).

Since the advent of patch clamping there has been a steady development of protocols and analysis techniques, many of which are now standard vocabulary to electrophysiologists, while others represent very specialised applications. However, two singular inventions have greatly enhanced the general versatility of patch clamping. These are perforated patch clamping and patch clamping in tissue, particularly brain slices. Ironically, both of these techniques overcome some of the possible disadvantages of patch clamping compared with 'good old' intracellular recording, namely washout of intracellular factors in whole-cell recording and the necessity to record from isolated cells. Since their introduction, both techniques have seen their own evolution and refinements.

Finally, although not dealt with in this book, it must be mentioned that many investigators have successfully combined patch clamping with other techniques to produce very elegant and powerful work. Examples of such symbioses are a combination of patch clamping with the recording of other physiological parameters such as calcium fluorescence or with the recording of single-cell contraction. The evolution of molecular biological techniques has also had a major impact on the world of electrophysiologists. It is possible to express ion channels and/or modulating factors in expression systems such as oocytes or cell lines and to find out how these exactly work by tweaking the expressed proteins one amino acid at a time. Molecular biology has helped electrophysiologists to identify new targets, link physiology with genetics, and more. It is a challenge for current bioscientists, more than ever before, to produce work that draws from different disciplines, either by teamwork or collaborations. The members of the team or collaboration must be able to communicate, which is something that electrophysiologists have had difficulty with owing to the cultural gap between chemistry-based biology and biophysics. Because this book is intended to introduce non-electrophysiologists to patch clamping, it is possible that it will help to bridge this gap from both sides.

2
Basic Theoretical Principles

This chapter deals with the combination of cell anatomical features and physical and chemical properties of the cell environment that contribute to the physiology of ion transport across the cell membrane. Starting from the relevant properties of the cell membrane and its intracellular and extracellular medium, we will move towards the introduction of some electronic principles and their application to this system, and end the chapter with electronic representations of cell membranes under a range of electrophysiological recording conditions. Understanding of the latter is absolutely essential for electrophysiological experimentation and data interpretation.

2.1 Introduction to Membrane Biology

2.1.1 The plasma membrane and its ionic environment

All living cells are enveloped by a plasma membrane that acts as a barrier between the cytoplasm and the extracellular space. The main constituents are phospholipids, which contain both lipophilic (fatty) and hydrophilic (polarised) residues (Figure 2.1).

In a watery environment phospholipids will arrange themselves spontaneously into structures where the lipophilic residues face each other. The arrangement found in cell membranes is a bilayer of phospholipids, forming a particularly effective barrier to charged molecules (Figure 2.2). The phospholipids are a dynamic if not fluid substrate in which other membrane constituents are embedded. These are mostly proteins with a variety of functions, most importantly communication (receptors and ion channels), structure (cytoskeletal anchors) and cellular homeostasis (e.g. ionic pumps, enzymes).

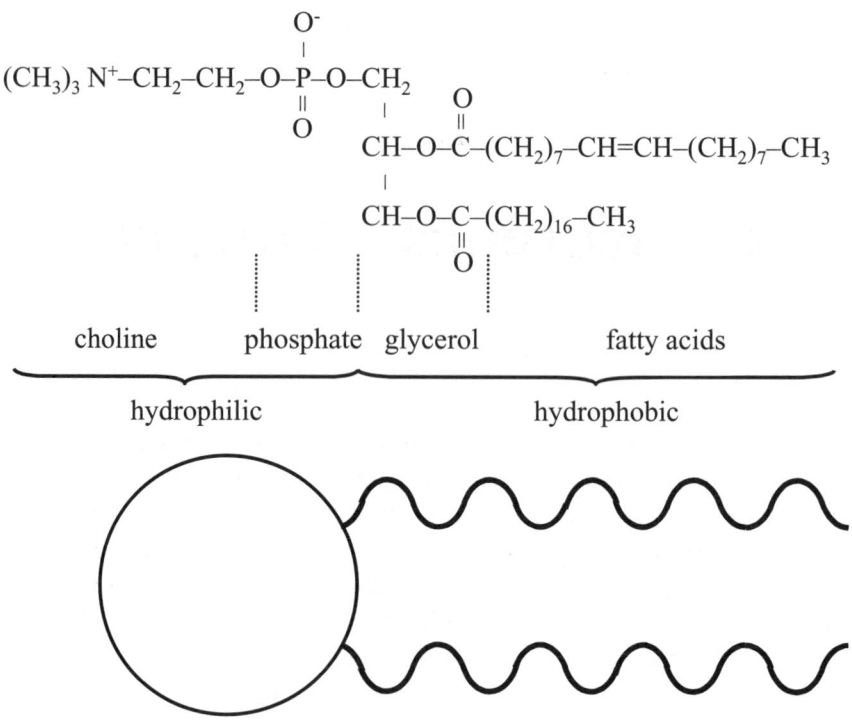

Figure 2.1 Phosphatidyl choline, which is a typical membrane phospholipid, has a polarised head and fatty tails. Phospholipids are often represented as shown at the bottom

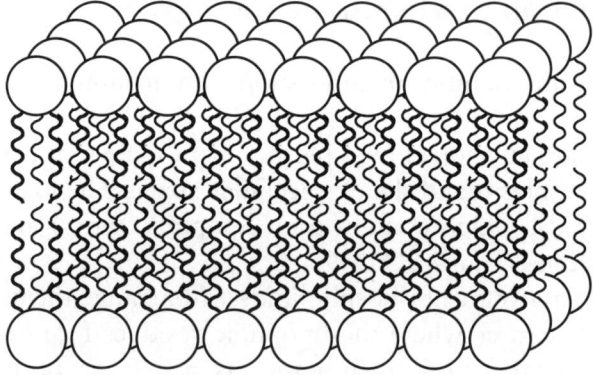

Figure 2.2 Bilayer arrangement of phospholipids in a watery environment

In many non-animal cells, the plasma membrane in turn is surrounded by a cell wall consisting mostly of mechanically strong material such as cellulose, which in conjunction with osmosis provides support and protection to the cell. In animals no such cell wall is present so the delicate plasma membrane must be protected from excessive osmotic forces by tight control of the osmolarity of the extracellular and intracellular media. Inorganic ions form the vast majority of particles in these media. As a result, the media possess a gross distribution of ions that is fairly constant over a broad range of cell types, organs and even animal species. An overview of ranges of concentrations of the principal ions in the extracellular and intracellular medium is presented in Table 2.1. Most striking are the large sodium and potassium ion concentration differences, on which many physiological processes depend. The calcium ion distribution in the intracellular and extracellular milieu, however, represents the greatest concentration difference: typically four orders of magnitude.

Table 2.1 Intracellular and extracellular distribution of the main ions found in animal fluids

Ion	Intracellular range (mM)	Extracellular range (mM)
Na^+	5–20	130–160
K^+	130–160	4–8
Ca^{2+}	50–1000 nM[a]	1.2–4
Mg^{2+}	10–20	1–5
Cl^-	1–60	100–140
HCO_3^-	1–3	20–30

[a] Given as nanomolar rather than millimolar.

2.1.2 Electrochemical gradients and the Nernst equation

The concentration differences between intra- and extracellular milieu introduced in the previous section result in concentration *gradients* for each ion across the plasma membrane. Concentration gradients induce the diffusion of particles from higher to lower concentration. Diffusion is the tendency of particles to spread equally over a space. A condition is that the particles have to move randomly, which they do in a fluid or gas unless the temperature is absolute zero. Thermodynamically, diffusion is a spontaneous process because it decreases order in a system, i.e. it increases entropy. Importantly, this implies also that diffusion releases energy. Walther Hermann Nernst (1864–1941) quantified this energy as

$$\Delta G = -RT \ln \frac{[ion]_o}{[ion]_i} \qquad (2.1)$$

where ΔG is the (Gibbs) energy to be released by the diffusion process, R is the universal gas constant (8.31 J mol^{-1} K^{-1}), T is the temperature in Kelvin and $[ion]_o$ and $[ion]_i$ are the extracellular and intracellular concentrations of the ion under consideration, respectively. Thus, if the plasma membrane is permeable to potassium ions, for example, then they will move from the cytoplasm to the extracellular space. The movement of positively charged potassium ions out of the cell will render the cell negatively charged, and that will start to attract the potassium ions back into the cell. The electrical energy responsible for this can be quantified as

$$\Delta G = -EzF \qquad (2.2)$$

where E is the potential across the membrane, z is the oxidation state of the ion under consideration and F is the Faraday constant (9.65×10^4 C mol^{-1}). If left unchecked, the diffusion of potassium from the cell will continue until the electrical force that this movement creates is equal in size but opposite to the diffusion energy. At this point there is no net movement of ions and the two forces are in equilibrium

$$-RT \ln \frac{[ion]_o}{[ion]_i} = -EzF \qquad (2.3)$$

The relation can be rearranged to provide the equation that describes the electrical potential at which a certain ion gradient is in equilibrium. This is the all-important Nernst equation

$$E = \frac{RT}{zF} \ln \frac{[ion]_o}{[ion]_i} \qquad (2.4)$$

where E is the equilibrium potential for the ion under consideration and is given the suffix for that ion, e.g., for potassium the potential will be named E_K.

2.1.3 Maintenance of ion gradients and the membrane potential

The flow of ions across the plasma membrane is facilitated by specialised proteins. It is useful to list here some terms that are used to refer to these proteins (not always correctly) in the literature:

2.1 INTRODUCTION TO MEMBRANE BIOLOGY

- *Ion channel*: a protein that facilitates ion diffusion across a membrane by forming an aqueous pore. Ion channels will be discussed further in Section 2.1.4.

- *Transporter*: the most general term for a membrane protein, which aids movement of molecules across the membrane *without* forming a pore. All proteins mentioned below are transporters.

- *Pump*: a transporter that transports molecules against their concentration gradient. The required energy is provided by the breakdown of ATP to ADP on the cytosolic side of the membrane. This process is named *active transport*, because it requires cellular metabolic activity.

- *Co-transporter or exchanger*: like a pump, a transporter that transports molecules against their concentration gradient, but here the energy is derived from the diffusion of other molecules, usually sodium ions. If the sodium ions move in the same direction as the transported molecules, then the transporter is a co-transporter, otherwise it is an exchanger. The ATP-dependent pumps subsequently restore the sodium ion gradient. For this reason, this form of transport is named *secondary active transport*.

- *Electrogenic*: if the transported molecules are charged, the activity of the transporter might result in a net influx or efflux of charge, influencing the membrane potential. The transporter is then said to be electrogenic.

Maintenance of the intracellular high potassium and low sodium ion concentrations is mainly performed by an Na^+/K^+ pump (Figure 2.3). As a pump, it utilises the energy from the breakdown of ATP to ADP to transport both sodium and potassium ions against their concentration gradients across the plasma membrane, hence it is also named Na^+/K^+-ATPase. The Na^+/K^+ pump is electrogenic because each cycle transports three sodium ions out of the cell and two potassium ions inward, a net outward flux rendering the cytoplasm more negatively charged. The importance of the sodium and potassium ion distribution to the cell is illustrated by the fact that the Na^+/K^+ pump is often the greatest single energy consumer in the cell. Perhaps more finely controlled is the intracellular calcium ion concentration. Calcium ions are pumped out of the cell by a Ca^{2+}/Na^+ exchanger. In addition, calcium ions are also stored by an ATP-dependent pump in parts of the endoplasmic reticulum,

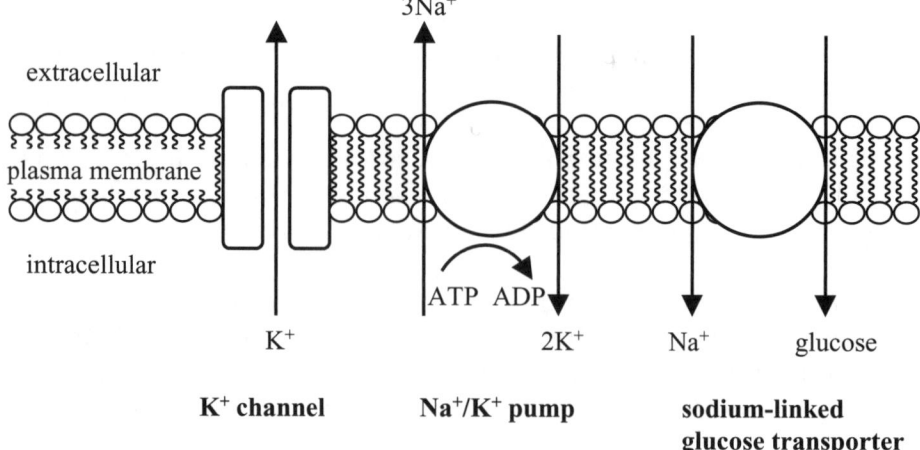

Figure 2.3 Examples of an ion channel, an electrogenic pump and a co-transporter, depicted by the symbols most often used for them

from where they can be mobilised to provide or enhance a cytosolic calcium response as part of a wide variety of signal transduction mechanisms. (A discussion of these mechanisms is outside the scope of this book.) Chloride ions, as the most important species of anions, were often considered to move passively in reaction to cation transport, but they also have transporters and ion channels and are therefore an often underrated physiological entity in themselves. In addition to inorganic ions, the majority of organic molecules in a cell have charged residues that contribute to the overall charge across the membrane, even though they are relatively stationary.

The charges on either side of the plasma membrane are not in balance; the inside of all cells at rest is more negatively charged than the outside, and the difference causes an electrical potential over the membrane. The reference for potentials in a cell system is, by convention, the extracellular medium, so resting membrane potentials are negative. How does this imbalance in ionic charge come about? At any moment, the relative permeability of the plasma membrane to each of the ions discussed above and their distribution on either side of the membrane will determine the membrane potential. For example, if the membrane is solely permeable to potassium ions, the situation is exactly as described earlier in discussing the Nernst equation. Potassium ions will diffuse out of the cell (along their concentration gradient), leaving the cytoplasm increasingly negatively charged. This will continue until the negative charge is large enough to keep the ions in the cell. The potential over the membrane then will be equal to the equilibrium potential for potassium ions E_K, which is typically -80 to -90 mV. In reality, the plasma membrane at rest will also be permeable to other ions, albeit to a much lesser extent, drawing the membrane potential towards their own equilibrium potentials. Because all other equilibrium potentials are more positive than E_K, the membrane potential will be more positive than E_K, typically -50 to -80 mV.

2.1.4 Ion channels

The previous section stated that the membrane potential is caused by a charge imbalance across the membrane, and that in most cells at rest the relative permeability of the membrane for potassium ions is dominant. Hence, the membrane potential is close to the potassium equilibrium potential, E_K. However, it was also pointed out that the phospholipid bilayer is an effective barrier for charged particles. How then is the permeability of the membrane for different ions established and controlled? Ion channels are a subset of proteins that span the plasma membrane. They possess several properties that make them very effective in controlling membrane permeability to small water-soluble molecules. These properties are:

1. An aqueous pore that connects the intracellular medium with the extracellular medium. This continuity provides the route by which diffusion across the membrane can take place. The pore is lined by mainly hydrophilic amino acid residues.

2. A gating mechanism that can close the pore. The gating mechanism is a change in protein conformation. The conformation change can be initiated by a range of factors that are dependent on the channel species. Classified according to gating factor, the three main ion channel species are listed below:

    - Voltage-dependent channels open or close depending on the membrane potential. These channels have voltage sensors: charged residues that shift position within the protein in response to membrane potential changes.

    - Ligand-gated channels open or close depending on the binding of an extracellular factor, such as a hormone or a neurotransmitter.

- Second messenger operated channels open or close in response to an intracellular factor, such as calcium ions or activated G protein subunits.

These three types of channel species are sometimes referred to as VOCs (voltage-operated channels, Figure 2.4), ROCs (receptor-operated channels) and SMOCs (second-messenger-operated channels), respectively. Note that these groups are not mutually exclusive. For example, there are calcium dependent potassium channels that are also voltage-dependent.

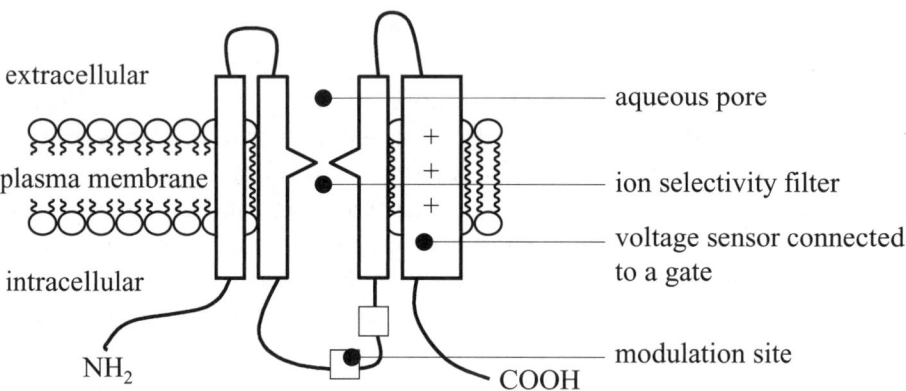

Figure 2.4 Diagram of a generic voltage-dependent channel

In addition, modulatory mechanisms can influence gating independent of the primary gating factor. Very common is phosphorylation of the protein on the cytosolic side, which provides a mechanism for fine tuning the channel's activity. Phosphorylation can enhance or reduce the function of the channel, depending on the phosphorylation site and the channel type. Another mechanism is differential expression. Because ion channels are proteins, they can be subject to varying levels of expression that can modulate channel function over longer time scales.

3. Selective permeability. Most ion channels show selectivity in that their pores are more permeable to some ions than to others. The mechanism of selective permeability is based on a combination of size of the ion (in its hydrated form) and its charge. Residues in the channel pore

lining interact with ions to form thermodynamic energy barriers that favour the passage of certain ions.

2.2 Electrical Properties of the Cell Membrane

To reach a point where the experimenter is able to interpret readily the signals recorded in an electrophysiological set-up, it is necessary to introduce some electronics terminology. However, compared with the electronics found in modern patch clamp amplifiers, the elements that make up the membrane in electrical terms are very simple indeed. Instead of these elements being introduced in their pure electronic forms, they will be introduced in terms of a membrane system so that the application is instantly obvious. At this point it is good to be reminded of the fact that electrical phenomena in biological systems are not mediated by electron movement, as in metals or semiconductors, but by the movement of ions in solution.

Units and symbols used in the next sections are summarised in Table 2.2 and Figure 2.5 for reference.

Table 2.2 Electrical parameters and units

Parameter	Symbol	Unit	Unit abbreviation
Potential/voltage	E	volt	V
Current	I	ampere	A
Resistance	R	ohm	Ω
Conductance	g	siemens	S
Capacitance	C	farad	F
Charge	Q	Coulomb	C

2.2.1 Driving force and membrane resistance

As explained in Section 2.1.1, the phospholipid bilayer is an effective barrier for the movement of charged particles. In contrast, the intra and extracellular media are watery salt solutions and very conductive to ions. Thus, the membrane forms an insulator between two conductors. The electrical insulation is not perfect: there are ion channels, transporters and there is some leakage. In other words, the resistance of the membrane to the movement of ions across it is finite. How can we quantify this?

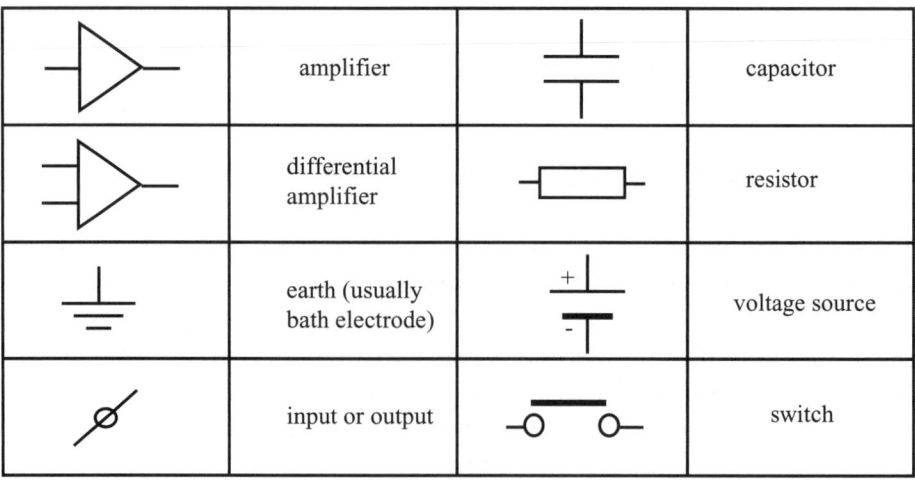

Figure 2.5 Electronic diagram symbols

In Section 2.1.3 we have seen that a potential difference exists between the inside and the outside of the cell, and that if this potential were completely dependent on potassium ions, the membrane potential would be at E_K. At this potential the force by which the electrical field is pulling potassium ions inward is exactly the same as the diffusion force pushing potassium ions outward. There would be no net movement of potassium ions across the membrane. However, if the membrane is not at E_K, then the membrane potential will not be in balance with the diffusion force and potassium ions will move across the membrane. In electronic terms, charge movement is expressed in current, i.e. charge movement per unit of time

$$I = \frac{dQ}{dt} \qquad (2.5)$$

where I is current in amperes (A) and dQ/dt is the change in charge in coulomb (C) over time. Electrical symbols and units are summarised in Table 2.2. Two things determine the size of the flow of ions (which will be called current hereafter): driving force and membrane resistance. Driving force is the difference between the equilibrium potential and the membrane potential E_m. It makes sense that the further the membrane potential is away from, in this case, the potassium equilibrium potential, the greater the imbalance is between electrical force and diffusion force, and therefore the greater the net flow of potassium ions. Thus, current is proportional to driving force (in volts). The current is limited by the resistance of the

2.2 ELECTRICAL PROPERTIES OF THE CELL MEMBRANE

membrane. If there are fewer potassium channels open, then fewer potassium ions can flow: current is inversely proportional to resistance. We can summarise the above by stating

$$I_K = \frac{E_m - E_K}{R_{m,K}} \quad (2.6)$$

where I_K is the potassium current, $E_m - E_K$ is the driving force for potassium and $R_{m,K}$ is the membrane resistance for potassium. Resistances are expressed in ohm (Ω). This equation is a special case of Ohm's law. If all ions pass through the membrane more or less equally, as in the case of leak current, then the equilibrium potential is 'neutral' so that any deviation from the resting membrane potential will result in a leak current of the size

$$I_{leak} = \frac{E_m - E_{rm}}{R_{m,leak}} \quad (2.7)$$

where $E_m - E_{rm}$ is the deviation from the resting membrane potential. We will see leak currents under voltage clamp conditions (Section 2.3.3).

2.2.2 Membrane capacitance

A membrane and the intracellular and extracellular media, from an electronic engineer's perspective, form a capacitor. A capacitor can store charge. In the case of a membrane, this occurs as follows: the membrane potential indicates that the intracellular side is more negative than the extracellular side, both sides exert an electromagnetic field across the membrane, which attracts charged particles, negative ions (anions) in the cytosol will be attracted to the positively charged outside and accumulate near the membrane, whereas positive ions (cations) will accumulate on the outside as shown in Figure 2.6.

This situation of captive ions can be regarded as an energy state, and the amount of charge stored can be calculated by

$$Q = E_m C \quad (2.8)$$

where Q is the charge stored, E_m is the potential difference across the membrane and C is the membrane capacitance (expressed in farad). This equation shows that capacitance is a measure of the capacity of the

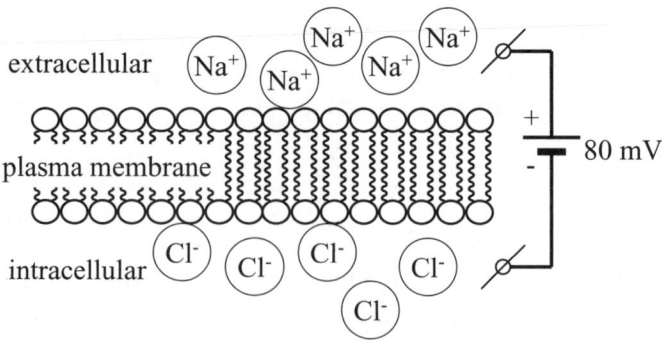

Figure 2.6 The membrane potential charges the membrane like a capacitor

membrane to store charge at a given potential. The physical dimensions of the membrane are important in determining the capacitance: the more membrane, the more charge can accumulate, therefore the capacitance is proportional to membrane surface area. The strength of the electromagnetic field that attracts the ions on either side of the membrane decreases with distance, therefore the capacitance is inversely proportional to membrane thickness. The latter variable turns out to be not very much of a variable at all: the phospholipid bilayer has a relatively constant thickness in all living cells, certainly in animal cells. Finally, the electromagnetic field is also dependent on the material separating the two conductors (the intracellular and extracellular media). The properties of the membrane pertaining to capacitance are summarised under the variable of dielectric constant or ε_r, and is also similar throughout living cells. In summary, capacitance can be described by

$$C = \frac{A\varepsilon_r}{d} \quad (2.9)$$

where A is the membrane area, ε_r is the dielectric constant for the membrane and d is the membrane thickness. Thus, a measurement of capacitance (see Section 4.2.1) provides a good estimation of the membrane surface area under investigation!

2.2.3 Consequences of membrane capacitance

The membrane capacitance plays an important role in both the physiological function of the plasma membrane and in the conduct of electro-

2.2 ELECTRICAL PROPERTIES OF THE CELL MEMBRANE

physiological experiments. In excitable cells, local changes in membrane potential, such as an action potential in an axon, travel along the membrane in a cascade of local circuit currents. If there are no ion channels involved in the propagation of the potential change, the axon will behave like a badly insulated electric cable, hence the associated phenomena are named cable properties. The signal will die out over distance due to the signal leaking across the membrane resistance. The membrane *capacitance* imparts a delay: any membrane potential change must first overcome a change in stored charge (either reduce it or increase it), in effect stretching membrane potential changes out over time. Thus, membrane capacitance is a limiting factor in action potential propagation speed.

Similarly, under experimental conditions such as voltage clamp (Section 2.3.3), where the membrane potential is controlled by the experimenter, a stepwise change in membrane potential will always come into effect at the cell membrane with some delay because of the membrane's capacitive properties, but there are ways of reducing this delay (see Section 4.2.1).

2.2.4 An electronic model of the plasma membrane

Given that an intact cell has a membrane potential, a resistance and a capacitance, the electronic representation of a cell looks like Figure 2.7.

In the case of an isolated patch of membrane, a membrane potential is not maintained so the voltage source would be left out of the diagram. The

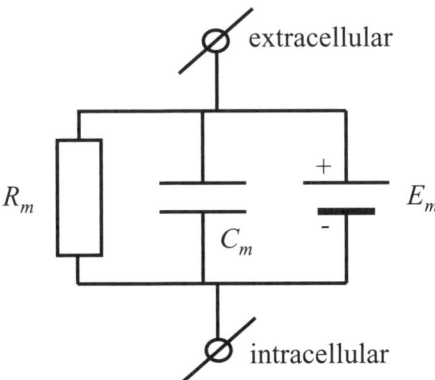

Figure 2.7 An electronic model of the plasma membrane of an intact cell. Membrane resistance R_m, capacitance C_m and membrane potential E_m are depicted by the appropriate electronic symbols

electronic representation of the cell membrane is the first very important step towards the construction of equivalent circuits, which greatly help to understand experimental configurations. These will be the subjects of the next section.

2.3 Recording Modes and their Equivalent Circuits

Broadly speaking, electrophysiological techniques to record ion fluxes across a membrane can be divided into *indirect* methods that employ extracellular electrodes, as in many non-invasive methods such as electro-encephalo/cardio/myo-grams, and *direct* methods that utilise micropipettes (see Section 3.3) to make contact with the cell of interest. The latter include intracellular recording techniques, where the pipette penetrates the cell, and patch clamp, where the pipette makes contact with the cell but does not penetrate. There is a small area of overlap between the direct and indirect methods, where micropipettes are used for extracellular recording of excitable cell activity. Indirect methods are not considered further here. In this section I introduce you to recording modes using microelectrodes. The properties of each recording mode will be discussed with the help of electronic representations or *equivalent circuits*. Most phenomena that you are ever likely to encounter working with micropipettes can be explained easily using these diagrams, so it is worth spending time studying them.

2.3.1 The basics of equivalent circuits

Equivalent circuits of electrophysiological experimental situations contain mainly resistors and capacitors. To facilitate the use of the circuits, relevant electronic principles pertaining to interactions between resistors and between resistors and capacitors are explained in this section.

Resistors

The current I through a resistor is proportional to the potential E across it, and inversely proportional to the resistance R. This is Ohm's law

$$I = \frac{E}{R} \tag{2.10}$$

2.3 RECORDING MODES AND THEIR EQUIVALENT CIRCUITS

Two resistors in series can be considered as one resistor with a resistance of the sum of the two original resistors, because any current has to pass both barriers (Figure 2.8).

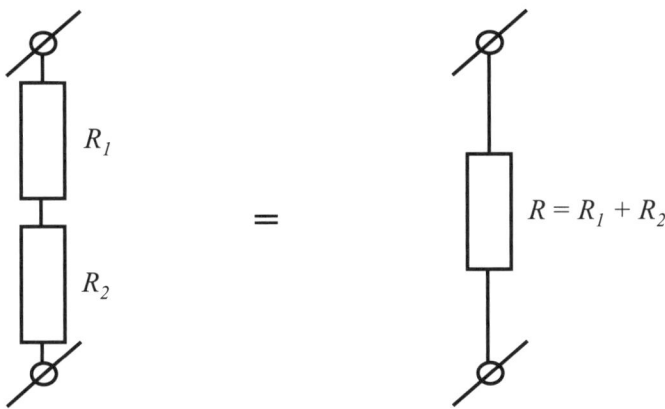

Figure 2.8 Resistors in series add up

A potential difference across two resistors in series will distribute itself over the resistors proportional to the resistance values. Each resistor represents a voltage drop, and all voltage drops in the circuit added up will be equal to the original potential. Figure 2.9 demonstrates an example: if the total resistance is 1050 MΩ, then the voltage drops over the two resistors are $1000/1050 \times 60 = 57.1$ mV for the 1 GΩ resistor and $50/1050 \times 60 = 2.9$ mV for the 50 MΩ resistor. This is in fact Kirchoff's voltage law. The current through each of the resistors is equal because it has to pass through both of them.

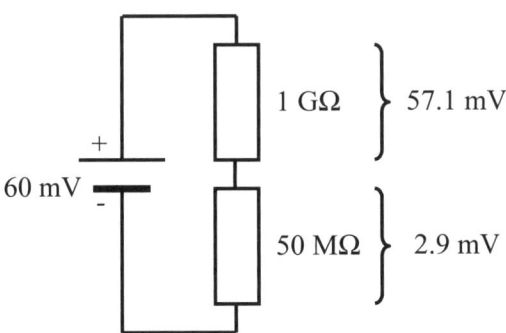

Figure 2.9 Voltage drops over resistors in series

In the case of resistors in parallel, the total resistance will be less than that of each individual resistor because there are now two pathways for current to flow. The total resistance is calculated using the reciprocal rule (Figure 2.10).

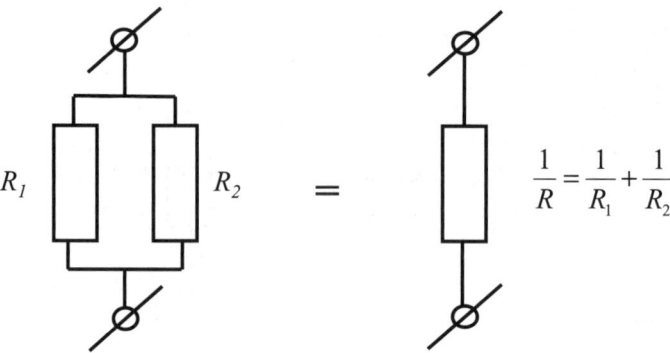

Figure 2.10 Parallel resistors form a lower total resistance than each individual resistor, because they form two separate possible pathways for current to flow. The total resistance is calculated using the reciprocal rule

If a potential is present, then both resistors will 'see' this potential, but the current through the resistors will be different and dependent on the individual resistance values. The total current is the sum of currents through all resistors. Figure 2.11 demonstrates an example: if the total resistance (using the reciprocal rule) is 333 MΩ, then the total current (using Ohm's law) is 180 pA, made up of the current through the 500 MΩ resistor (120 pA) plus the current through the 1 GΩ resistor (60 pA). This is a guise of Kirchoff's current law.

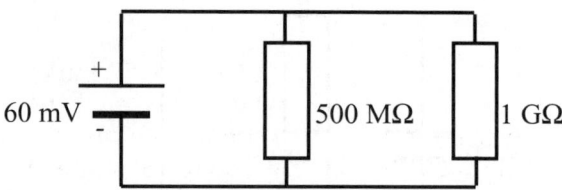

Figure 2.11 Currents through parallel resistors add up to the total current

In practice, Kirchoff's laws imply that:

2.3 RECORDING MODES AND THEIR EQUIVALENT CIRCUITS

- Voltage should be measured over a high resistance, other resistances *in series* being minimised as much as possible.

- Current is drawn by *parallel* resistors to the recording circuit which can cause *short-circuiting*, so they should be made as large as possible.

Examples of these are discussed in Section 2.3.4 and elsewhere.

Resistors and capacitors

A resistor and a capacitor in series form an *RC* circuit, which plays an important role in many equivalent circuits. The speed of charging and discharging a capacitor, such as the plasma membrane, upon a change in potential depends on the resistor(s) in series with it. If a voltage is applied to an *RC* circuit, the voltage over the capacitor will build up exponentially. The circuit resembles Figure 2.12 and its behaviour is demonstrated in Figure 2.13.

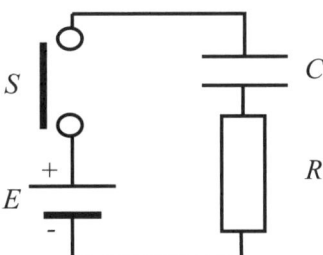

Figure 2.12 Circuit to demonstrate the behaviour of *RC* couples, such as a plasma membrane and a patch pipette (see Section 2.3.4 and elsewhere). Closing the switch 'S' will change the potential over the *RC* circuit

Closing of the switch (Figure 2.13, at the arrow) changes the potential over the whole circuit instantaneously (top graph), but only gradually over the capacitor (middle graph), following an exponential curve. The current initially will surge and then decline exponentially (bottom graph). Quantitatively, the exponential curve that describes the capacitor potential over time t is the time elapsed from closing of the switch

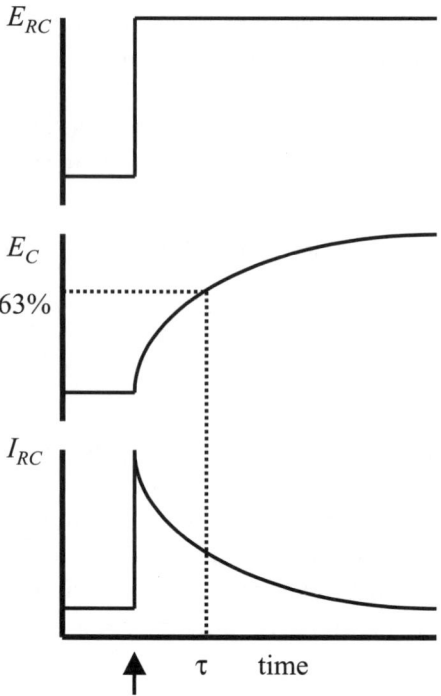

Figure 2.13 The effect of a voltage change on an RC circuit. Note the bottom graph; this will reappear in whole-cell voltage clamp experiments (Chapter 4.2.1)

$$E(t) = E(1 - e^{-t/\tau}) \qquad (2.11)$$

where τ is the time constant of the RC circuit. The equation shows that at time $t = \tau$ the capacitor is loaded to $(1 - e^{-1})$ of its maximum. This equates to 63 per cent.

The resistance and the capacitance are both linearly proportional to the time constant τ

$$\tau = RC \qquad (2.12)$$

The concepts of series and parallel resistors and RC circuits will help in determining the attributes that elements in an experimental situation need to possess for the experiments to be successful or even meaningful. In the following sections this will be demonstrated.

2.3.2 Intracellular recording

Although this volume is about patch clamping, it is useful to consider intracellular recording first as a simple example of an electrophysiological recording configuration. Intracellular recording involves puncturing of the plasma membrane. With one electrode making direct contact with the cytosol and another electrode present in the bath, the two electrodes are on either side of the membrane and so allow direct measurement of the membrane potential. As a consequence of the need for penetration, it requires relatively sharp glass micropipettes to reduce the damage to the cell. The tip of a pipette for intracellular recording has a diameter in the order of tens of nanometres. The small tip of the pipette limits both the electrical conductivity of the pipette (resulting in a relatively high pipette resistance) and the washout of cytoplasm by pipette fluid. The equivalent circuit for this configuration is presented in Figure 2.14 and will be considered below.

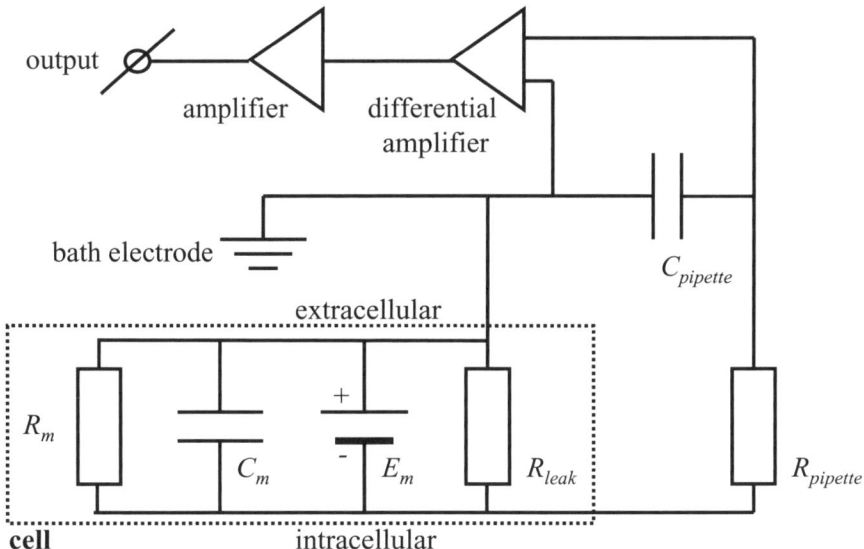

Figure 2.14 Equivalent circuit for an intracellular recording configuration. In addition to the membrane resistance, capacitance and potential, a leak R_{leak} is introduced due to rupturing of the membrane by the pipette

In intracellular recording, the cell is penetrated by a glass pipette in order to make an electrical circuit between the electrode in the micropipette and the cytoplasm. In this way, the potential difference between the bath electrode and the electrode in the pipette directly reflects the membrane potential. There are some important considerations here, the discussion of which follows a pattern that can be applied to all equivalent circuits:

1. *Pipette resistance*: The small size of the tip of the micropipette creates a resistance. The resistance is usually minimised by using a highly conductive solution (2–3 M KCl) to fill the pipette and form a connection with the metal junction that leads to the probe. This can be done relatively safely because, owing to the small tip size, the leakage from the pipette into the cell is minimal (unlike with patch electrodes!). Microelectrodes for intracellular recording have resistances of 15–150 MΩ, where generally sharper electrodes (higher resistance) are necessary for smaller cells. A high pipette resistance does not have to be a problem. The membrane potential, according to the equivalent circuit of Figure 2.14, is distributed over the pipette resistance and the input of the differential amplifier. The differential amplifier records the voltage difference between two inputs. In practice the differential amplifier is housed separately from the main amplifier (see Section 3.4.2) and is usually referred to as a 'probe', although strictly speaking the 'probe' is the physical housing that contains the differential amplifier and other circuits. According to Kirchhoff's voltage law, the greatest voltage drop in a series circuit will be over the highest resistance, so if the input resistance of the probe is very high compared with the pipette resistance, then the probe will 'see' most of the membrane potential. Modern probe resistances are very high indeed (>1 GΩ), as illustrated in Figure 2.15.

   For a realistic example:

   $E_m = -60$ mV

   $R_{pipette} = 50$ MΩ

   $R_{probe} = >1$ GΩ

   The total resistance is $1000 + 50 = 1050$ MΩ, so the probe 'sees' at least $1000/1050 \times (-60) = -57.1$ mV.

2. *Pipette capacitance*: the pipette glass is an insulator between the bath solution and the pipette solution, i.e. it forms a capacitor (Figure 2.10). As seen in Section 2.2.2, capacitors will delay potential changes and this situation is no exception. In practice, fast membrane potential

2.3 RECORDING MODES AND THEIR EQUIVALENT CIRCUITS

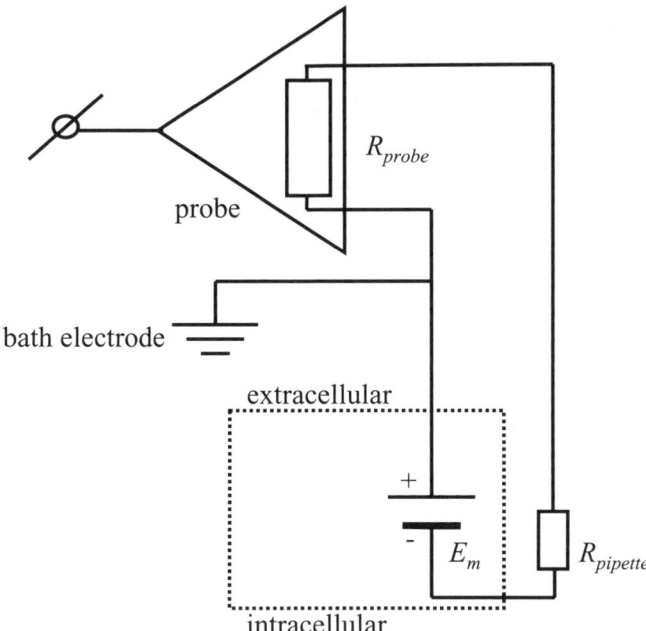

Figure 2.15 Simplified equivalent circuit for intracellular recording, showing the effect of pipette resistance on the potential recorded at the probe inputs

changes such as action potentials can be distorted by this effect and many amplifiers have circuitry built in that allows for the introduction of negative capacity to counteract it (see also Section 3.4.2). The circuitry can be set by the experimenter by means of a 'pipette capacitance' controller, and overcompensation can create problems. When this happens, the circuit becomes *hyper*sensitive to potential changes and will oscillate readily. This is a problem with all capacitance compensation systems.

3. *The leak resistance*: R_{leak} was introduced because of the damage inflicted by the microelectrode on the plasma membrane, creating effectively a short circuit from the cytosol to ground (the bath, see Figure 2.16). If this resistance is low (i.e., the hole in the membrane is large), then there will be a considerable load on the membrane potential that the cell might not be able to sustain. The membrane potential will become less negative and the cell will die. It is imperative that the damage is kept to a minimum, thus maximising R_{leak}. In practice this is done by very fast, well-controlled movement of the

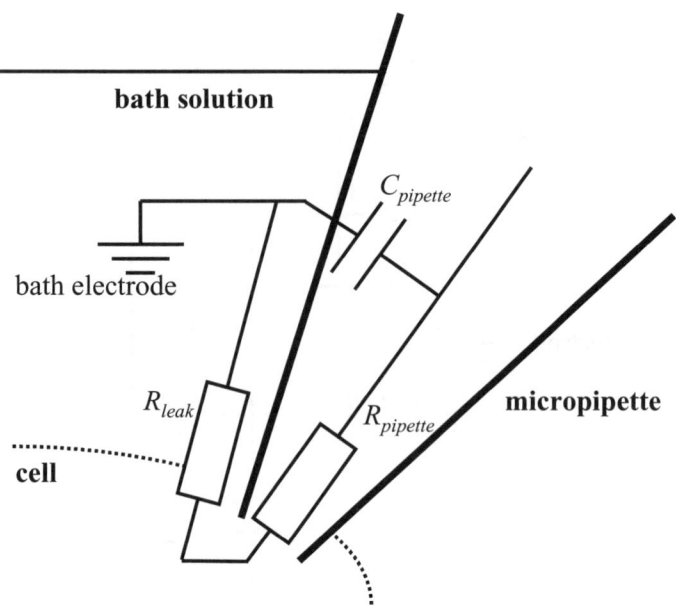

Figure 2.16 A closer look at the cell–pipette system

pipette, e.g. by using a piezoelectric stepmotor that translates a potential into motion (the reverse happens in a piezoelectric gaslighter).

4. *Current injection*: membrane resistance and capacitance can be measured by injecting a small current into the cell through the pipette (the additional circuitry for this is not shown in the equivalent circuit in Figure 2.14). The resulting voltage deflection can now used to estimate R_m and C_m. The relevant circuit looks like Figure 2.17.

The voltage deflection caused by the injected current is dependent on the resistance of the circuit, according to Ohm's law

$$E = IR \qquad (2.13)$$

When the current is constant, the voltage deflection is proportional to the resistance, which is made up of $R_{pipette}$ in series with the parallel resistors R_{leak} and R_m. Resistance $R_{pipette}$ is known because it is good practice to measure voltage deflection upon current injection before entering the cell, and R_{leak} is hopefully very high. Thus, the membrane resistance can be calculated as

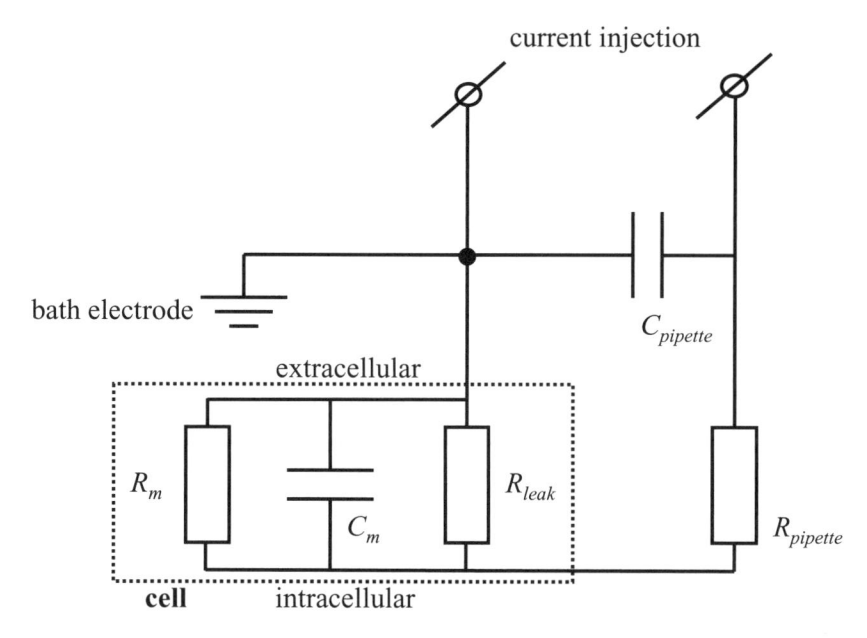

2.3 RECORDING MODES AND THEIR EQUIVALENT CIRCUITS

Figure 2.17 Elements that play a role in current injection in an intracellular recording configuration

$$R_m = \frac{E}{I} - R_{pipette} \tag{2.14}$$

An example of a current injection experiment is shown in Figure 2.18.

An injection of 20 pA results in a 10 mV voltage deflection, so the resistance of the system is 10 mV/20 pA = 500 MΩ. We assume that the current injection was also performed when the pipette was in the bath but not touching the cell yet, and that the resistance found was 50 MΩ. The membrane resistance together with the leak resistance is then 500 − 50 = 450 MΩ. The leak resistance must be very high indeed to obtain such a value, so this recording looks sound. (This is also apparent from the stability of the membrane potential in the potential graph.)

The membrane capacitance can be determined with the same experiment. There is an RC circuit in the total circuitry consisting of the pipette resistance (which was 50 MΩ) and the membrane capacitance. Note that the pipette capacitance is neglected, which is valid if it is small compared with the membrane capacitance (see Figure 2.17). The typical effect of the RC can be seen in the potential graph, which shows a sluggish reaction to the instantaneous current injection. The time constant τ of an RC circuit is

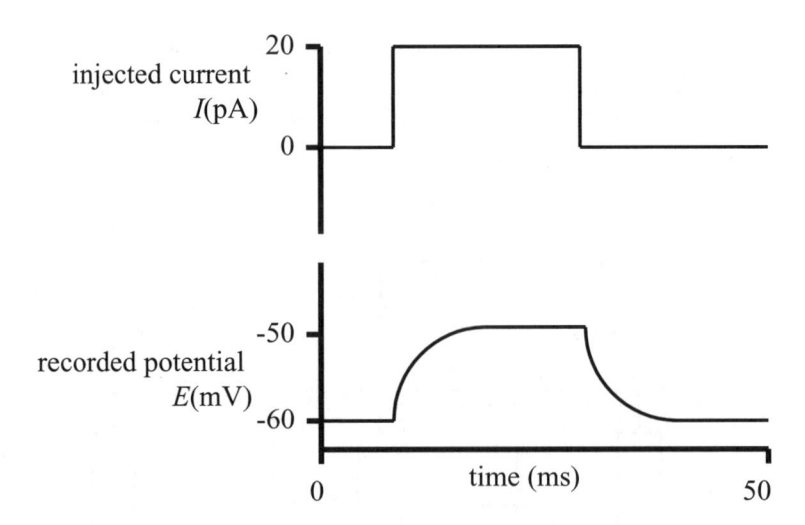

Figure 2.18 Response (bottom graph) to an injection of a small current (top graph) in an intracellular recording configuration. Some realistic values are shown as examples

defined as the time it takes for the response to reach 63 per cent of its maximum (as discussed in Section 2.3.1). In this case τ is 3 ms. We can now calculate C_m. Time constant $\tau = RC$, so $C = \tau/R$. The membrane capacitance is 60 pF. This value gives a good idea of the size of the cell, not taking into account space clamp (Section 5.1).

With the help of the equivalent circuit diagram, it is possible to build up an experimental routine very quickly in which the data can be analysed and interpreted easily, and the cause of anything untoward happening can be pinpointed. We will now focus on patch clamp configurations, starting with the central concept of voltage clamp.

2.3.3 Voltage clamp and current clamp

The activity of ion channels can be recorded in the form of membrane potential changes that reflect the membrane response to ions leaving or entering the cell. This is usually done in intracellular recording. Because changes in membrane potential are often the physiological effector of the ion channel activity (e.g. an action potential in a neurone), this parameter is often an important indicator of physiological significance of the activity. However, detailed study of the ion channel behaviour is very difficult

2.3 RECORDING MODES AND THEIR EQUIVALENT CIRCUITS

because membrane potential changes feed back to ion channel function in at least three independent ways:

1. The ion flow through the channels changes because of the changing driving force (as discussed in Section 2.2.1).

2. Many ion channels are voltage dependent, i.e. have gating mechanisms which respond to membrane potential changes.

3. Some ion channels have the ion flow restricted at certain membrane potentials by cytosolic factors. For example, magnesium ions can block NMDA channels and inward rectifying potassium channels when the membrane potential forces them into the channel pore.

It is therefore often desirable to control the membrane potential and record the membrane current directly. This situation is called voltage clamp. (Note that the different types of 'clamp' in electrophysiology can be confusing and can relate to totally different things.) Voltage clamp occurs usually through an electronic feedback system where a measured potential is compared with a potential set by the experimenter (holding potential). Any deviation of the recorded potential from the holding potential is instantly corrected by compensatory current injection. This current is then an accurate representation (but opposite in sign) of the ionic current over the membrane under investigation (Figure 2.19).

Because voltage measurement and current injection use the same pipette, and the current feedback must be instantaneous, a sharp electrode as used for intracellular recording is unsuitable. With low-resistance patch electrodes, however, voltage clamp is very feasible. Most patch clamp experiments are voltage clamp experiments. If the voltage is not clamped and a fixed amount of current is injected, then this is called current clamp. Current clamp does not require a feedback system and small currents can be injected through sharp intracellular recording pipettes, as seen in the previous section. In patch clamping, current clamp is used most often in either of two situations:

1. At the start of an experiment using the whole-cell configuration (Section 2.3.6), at the point where access to the cytoplasm is gained, a brief switch to current clamp with no injected current ($I = 0$) will effectively measure the membrane potential. This is a useful check on the quality of the patch and the cell.

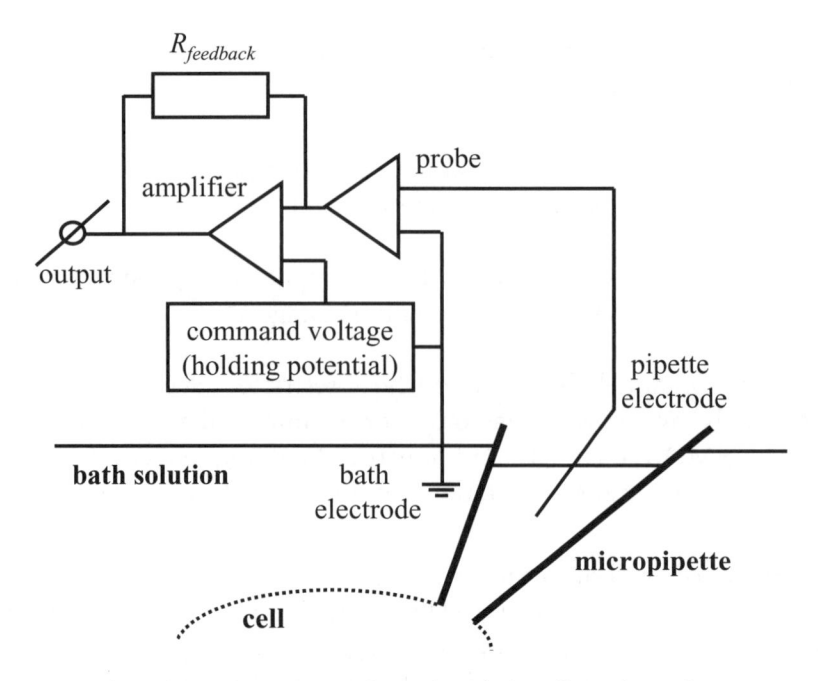

Figure 2.19 The voltage clamp principle in the whole-cell patch configuration. A command voltage is set by the experimenter and continuously compared with the measured potential. Any difference is instantaneously compensated by current injection. Whole-cell patch clamp is discussed in detail in Sections 2.3.6 and 4.2

2. Recording of synaptic potentials can be done in current clamp. Neuronal networks (in culture or in slices) often do not allow voltage clamp because of the cells' space clamp properties. Current clamp can then still be used to monitor the membrane potential, with the advantage over intracellular recording that the intracellular ionic composition can be controlled. These situations are discussed in more detail in Chapter 5.

In voltage clamp, the command voltage set by the experimenter can be made to vary over time from simple stepwise changes to complicated waveforms to study specific channel properties. These protocols are discussed in Section 5.2.

For completeness it should be mentioned that ways have been devised to use voltage clamp with sharp electrodes. These involve the use of two electrodes: one to record voltage and another to inject current. The method is aptly named dual-electrode voltage clamp. The method ob-

2.3 RECORDING MODES AND THEIR EQUIVALENT CIRCUITS

viously requires big cells and is often used for oocyte work (Figure 2.20). A second method, single-electrode voltage clamp, uses electronics to switch quickly between measuring potential and injecting current through the same micropipette. To obtain a good temporal resolution, switching needs to be quick (2–20 kHz), which results in very twitchy electronics (easily going into uncontrolled oscillation), and these experiments are notoriously hard to perform (Figure 2.21). The rationale for these methods is the need for voltage clamp of cells that cannot be patched (such as those in most tissues) and/or in situations where the cytoplasm must not be disturbed. The latter reason is now of less importance with the advent of perforated patch clamp (Section 4.2.2).

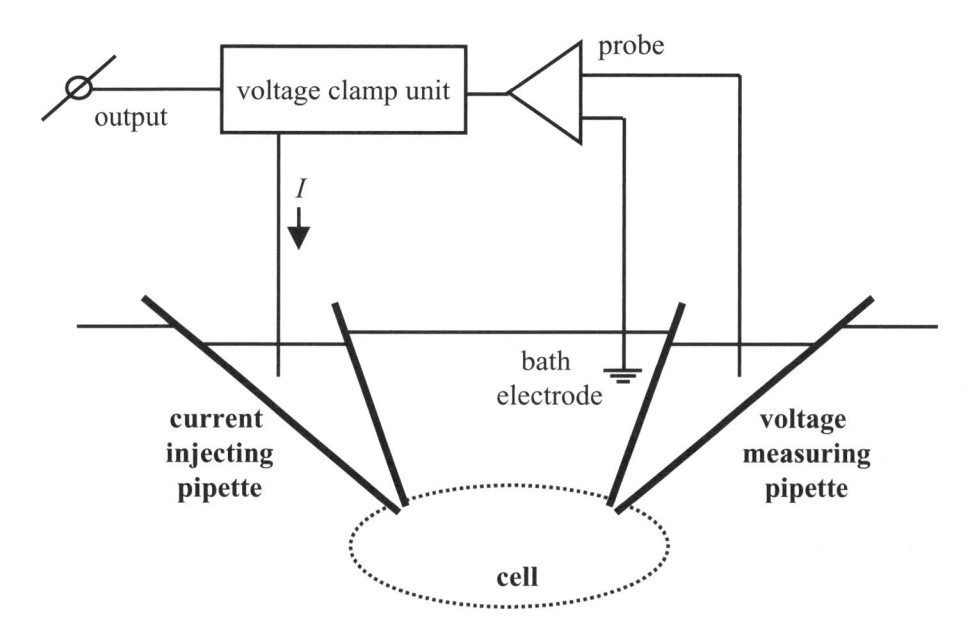

Figure 2.20 The dual-electrode voltage clamp principle

2.3.4 Introduction to patch clamp configurations

One of the features of patch clamp that makes the method so powerful is that it can be used in different guises, which enables the experimenter to:

- study ion channels at different levels; either whole-cell (activity of all ion channels added up) or individual ion channels;

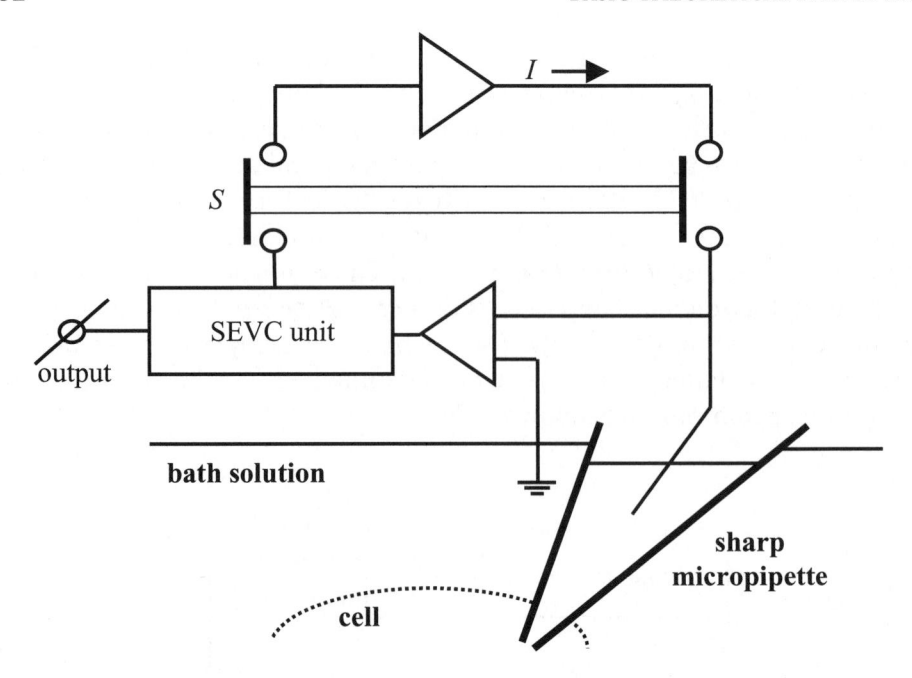

Figure 2.21 The single-electrode voltage clamp (SEVC) principle. The operation of the SEVC unit is rapidly switched between measuring potential and injection current. The switching frequency can be up to 20 kHz

- manipulate easily the fluid on the extracellular or the intracellular side of the membrane *during* a recording.

The various patch clamp configurations are introduced briefly in this section, and subsequently their equivalent circuits will be discussed.

Cell-attached patch mode

The simplest patch clamp configuration (in terms of physical manipulation) is cell-attached patch mode. Every patch clamp experiment starts with this situation. The micropipette, unlike intracellular recording, is positioned against the cell membrane where the glass makes a very strong connection, resulting in a tight (high resistance) seal. Ion channel activity in the tiny patch of membrane surrounded by the tip of the pipette can now be studied, as we shall see in the next section. The cell-attached patch mode is therefore a *single-channel configuration* (Figure 2.22).

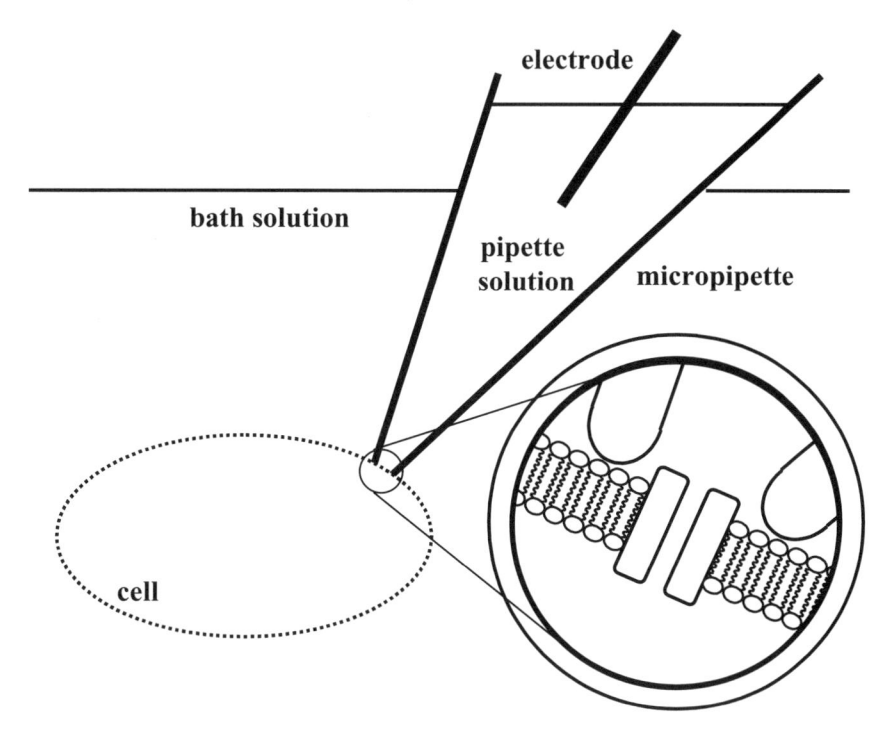

2.3 RECORDING MODES AND THEIR EQUIVALENT CIRCUITS

Figure 2.22 The cell-attached patch configuration

This configuration leaves the cell intact, and is therefore the most 'physiological' configuration to study single channels and the simplest to obtain. However, it does not allow easy manipulation of the media on either side of the membrane, and control over the potential over the patch (voltage clamp) has uncertainties because the membrane potential, which cannot be measured directly in this configuration, is still intact. An overview of advantages and disadvantages of patch clamp configurations is presented in Table 2.3. Because the pipette is on the extracellular side of the membrane, it is usually filled with bathing solution.

Whole-cell mode

If the patch of membrane under the pipette tip in cell-attached patch mode is ruptured, then the pipette solution and the electrode make direct electrical contact with the cytoplasm. The situation is very similar (but not quite identical) to intracellular recording: the patch electrode, electrically, is on one side of the plasma membrane and the ground electrode is on the

Table 2.3 Summary of advantages and disadvantages of patch clamp configurations

Configuration	Use	Advantages	Disadvantages
Cell-attached patch	Single-channel recording	Cytosolic side intact (physiological) Easy to obtain	Exact patch potential unknown No easy superfusion possible
Outside-out excised patch	Single-channel recording	Extracellular side can be superfused Cytosolic environment is controlled	Washout of cytosolic factors Disruption of cytoskeletal structure
Inside-out excised patch	Single-channel recording	Cytosolic side can be superfused Extracellular environment is controlled	Bath solution must be replaced by intracellular solution Relatively difficult to obtain Disruption of cytoskeletal structure
Whole-cell	Macro-current recording	Quick assertion of ion channel populations Cytosolic environment is controlled Extracellular side can be superfused	Washout of cytosolic factors
Perforated patch	Macro-current recording	Cytosolic ionic environment is controlled No washout of organic factors Extracellular side can be superfused	No control over organic cytosolic factors Relatively difficult to obtain

other, therefore the membrane potential can be recorded directly. The main difference with intracellular recording using 'sharp electrodes' is that the pipette resistance is relatively low, so larger currents can be injected quicker, allowing voltage clamp. A diagram of a whole-cell configuration is shown in Figure 2.19. A patch pipette tip is sufficiently wide to allow washout of the cytoplasm by the pipette-filling solution. Because the volume of the cell is negligible compared with that of the patch pipette, the composition of the intracellular fluid can be considered equal to that of the pipette-filling solution. If a 'physiological' situation is desired, then the pipette is filled with a solution of ionic composition that resembles the cytoplasm as closely as possible.

Washout can be a blessing or a curse: the experimenter can manipulate the intracellular milieu, but unknown cytosolic factors relevant to the subject of study can be unwittingly washed out. To avoid the latter, intracellular recording was used, but in many cases it is now possible to apply *perforated patch clamp*, where electrical contact with the cytosol is established by adding a membrane-perforating agent to the pipette solution. The agent (nystatin or amphoteracin B) perforates the membrane so that only small molecules such as ions can pass through, leaving the cytoplasm's organic composition largely intact. Ideally, the equivalent circuit for the perforated patch configuration is equal to that for the whole-cell configuration, differing only quantitatively in the size of the access resistance (Section 2.3.6). The procedure for perforated patch clamp is discussed in Section 4.2.2.

The outside-out and inside-out excised patch configurations

There are two single-channel configurations that do away with the cell altogether by excising a patch of membrane from the cell and studying it in isolation. This provides the experimenter with ultimate control over the environment of the patch and any ion channels in it, both electrically and chemically. 'Outside' and 'inside' refer here to the extracellular and intracellular side of the membrane, respectively, and 'out' refers to the bath. The outside-out patch is obtained by simply pulling away the patch pipette from a whole-cell configuration. The membrane will eventually break and, owing to the properties of the phospholipids, fold back on itself into a patch covering the pipette (Figure 2.23, left-hand panel). Note that for a 'physiological' situation the pipette solution should resemble the intracellular ionic environment because it is facing the intracellular side of the membrane. Outside-out patches can be used to study the effects of

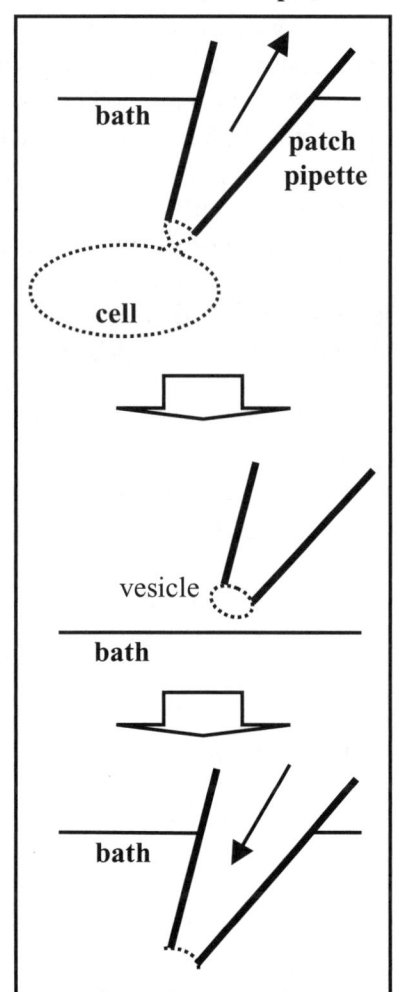

Figure 2.23 The excised patch configurations

extracellular factors on the channels, because the bath composition can be altered easily during recording.

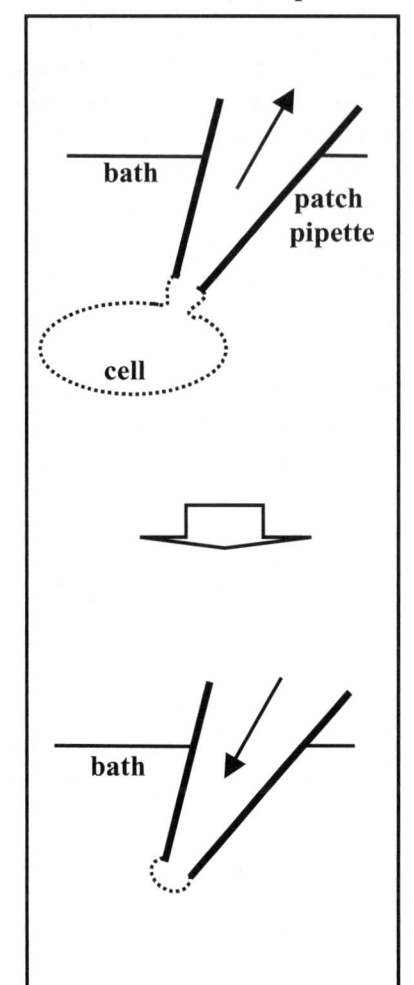

The inside-out excised patch is obtained from a cell-attached patch configuration, where again the pipette is pulled away. The result is now a vesicle attached to the pipette tip. The vesicle can be destroyed by exposure to air, i.e. the pipette is briefly lifted above the bath. This leaves a patch with the cytosolic side facing the bath (Figure 2.23, right-hand panel). Inside-out patches are ideal for studying the effects of cytosolic factors on

2.3 RECORDING MODES AND THEIR EQUIVALENT CIRCUITS

channels. It is clear that inside-out patches are much harder to work with than outside-out patches. To obtain them involves an extra step (destruction of the vesicle), and the bath solution must be replaced with intracellular solution in each experiment (see also Section 4.3.3).

2.3.5 The equivalent circuit for the cell-attached patch configuration

The resistors present in this circuit (Figure 2.24) differ from the intracellular recording situation only in that the pipette resistance is relatively low, but the resistance of the patch of membrane is very high. We have seen that the highest resistance in a series circuit determines the current flow, therefore if the patch resistance R_{patch} is high compared with the resistance of the rest of the cell (R_m) and the pipette resistance ($R_{pipette}$), then the circuit effectively monitors current flow through the patch and any ion channels in it.

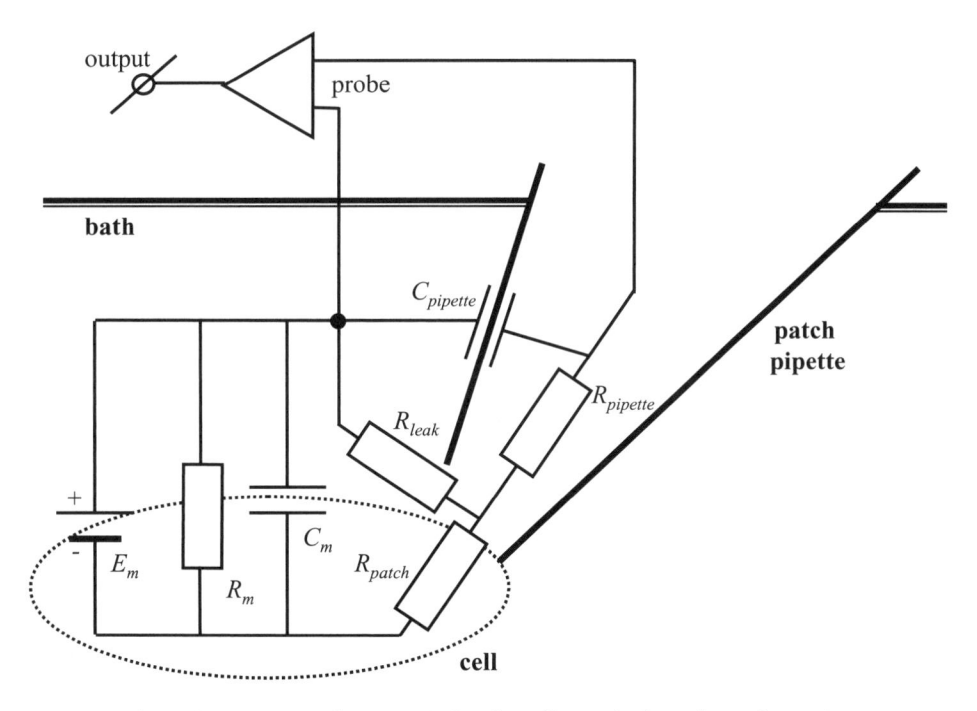

Figure 2.24 Equivalent circuit for the cell-attached patch configuration

There is one parallel resistor in the circuit, with the potential of short-circuiting, i.e. draining away current. Leak resistance R_{leak} represents the quality of the seal between the glass of the micropipette and the membrane. If the seal is good, then R_{leak} is very high and no significant current will leak away. How high is very high? The current through R_{patch} has two pathways to ground (the bath) — through the probe and through R_{leak} — therefore R_{leak} should be much higher than the current input resistance of the probe. In addition, good voltage clamp of R_{patch} can take place only if the parallel R_{leak} is not too low, and a low R_{leak} also significantly increases noise (noise will be dealt with further in Section 3.4.3). In practice, this means that R_{leak} should be better than 10 GΩ!

The relevant capacitances in the circuit are the pipette capacitance $C_{pipette}$ and the capacitance of the patch of membrane. The latter is very small indeed and not drawn in the equivalent circuit of Figure 2.24. The whole-cell capacitance C_m is not so important here because the whole-cell membrane resistance R_m is usually so much smaller than R_{patch} that R_m effectively short-circuits C_m, i.e. the circuit does not 'see' C_m. The pipette capacitance $C_{pipette}$, although small, must be well compensated for because the high magnification and fast time scale often required in single-channel recording greatly exacerbate capacitive artefacts.

Voltage clamp in the cell-attached patch configuration requires careful consideration because of the presence of the cell with its membrane potential. It must always be remembered that the voltage clamp circuitry clamps the potential between the two electrodes (the pipette and the bath electrode) and cannot distinguish between elements that the experimenter might or might not be interested in. From the perspective of the patch, the clamped potential is relative to the membrane potential. This is illustrated in Figure 2.25.

It is obvious that cell-attached patch experimentation and data analysis are hindered by the fact that the membrane potential is unknown. There are several ways of dealing with this:

- An average membrane potential, obtained using intracellular recording or whole-cell patch clamp, is assumed to be the membrane potential of the cell under investigation.

- At the end of the experiment, the patch is broken and the potential is recorded (under zero current clamp conditions, see Section 4.2.1). This must happen very fast because the intracellular ionic environment is disturbed through washout with pipette solution (usually extracellular solution).

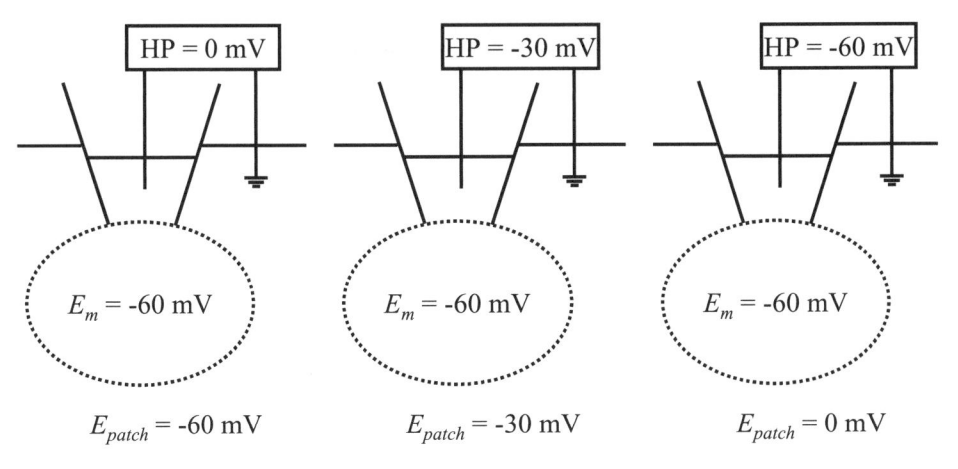

2.3 RECORDING MODES AND THEIR EQUIVALENT CIRCUITS

HP = 0 mV	HP = -30 mV	HP = -60 mV
E_m = -60 mV	E_m = -60 mV	E_m = -60 mV
E_{patch} = -60 mV	E_{patch} = -30 mV	E_{patch} = 0 mV

Figure 2.25 Voltage clamp in the cell-attached patch configuration. To obtain the actual potential over the patch, the command voltage (holding potential, HP) must be subtracted from the cell membrane potential

- Current–voltage characteristics of the ion channels studied can, in combination with other types of single channel data, sometimes provide clues to the membrane potential (more on this in Section 6.1.2). This is a retrograde method and does not help the experimenter during the experiment.

These methods are all indirect and their results suffer from variability between cells and/or any temporal changes in membrane potential. However, in many circumstances this will be outweighed by the ease of obtaining the configuration and the 'physiological' environment that the channels are exposed to.

2.3.6 The equivalent circuit for the whole-cell configuration

The whole-cell configuration is obtained by disrupting the patch of membrane under the pipette in a cell-attached patch situation. In electronic terms, this implies that R_{patch} becomes very low and, as such, is usually renamed the access resistance R_{access}. The membrane potential is disrupted as the integrity of the plasma membrane is lost and the pipette electrode has direct electrical contact with the cytoplasm. As a result, the equivalent circuit for the whole-cell configuration (Figure 2.26) is slightly simpler

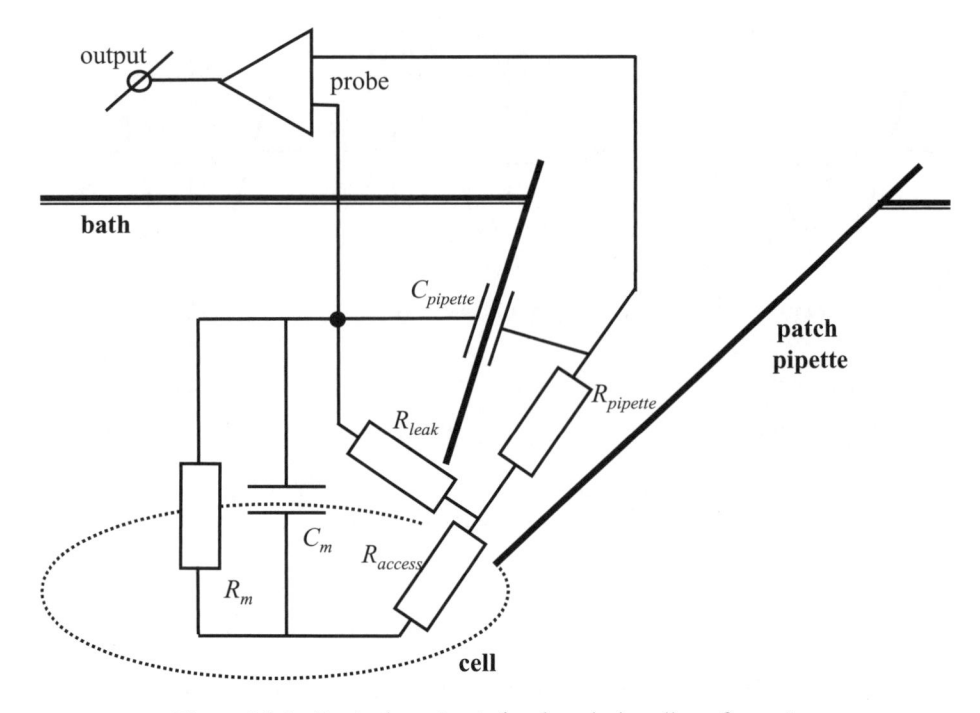

Figure 2.26 Equivalent circuit for the whole-cell configuration

than that for the cell-attached patch configuration, and much easier to apply.

The series circuit consists of the pipette resistance $R_{pipette}$, the access resistance R_{access} and the membrane resistance R_m. The latter is the largest (current-limiting) resistor, so this configuration allows the observation of currents through R_m. Because these currents are the sum of currents through all activated single channels of the cell, they are named whole-cell currents or macro-currents. Parallel to the circuit is the leak resistance R_{leak}, which should be as high as possible to minimise short-circuiting of the membrane current.

The membrane capacitance plays an important role in whole-cell recording, mainly because it affects the voltage clamp time characteristics. Any change in holding potential will be delayed because R_{access} and $R_{pipette}$ in series with C_m form a significant RC circuit (see Section 2.2.3). The sum of R_{access} and $R_{pipette}$ is sometimes referred to as series resistance. Methods to quantify and minimise the effects of this RC circuit are discussed in Section 4.2.1.

2.3.7 The equivalent circuit for the excised patch configurations

Excised patch configurations (introduced in Section 2.3.4) are represented by the simplest equivalent circuits. In principle, the diagram for whole-cell configuration can be applied, minus the membrane capacitance (which is very small indeed for a patch) and the access resistance. The result is shown in Figure 2.27. Excised patches are used for single-channel studies, which impose tight limits on the quality of the patch seal (as indicated by a very high leak resistance) and the maximum size of the pipette capacity.

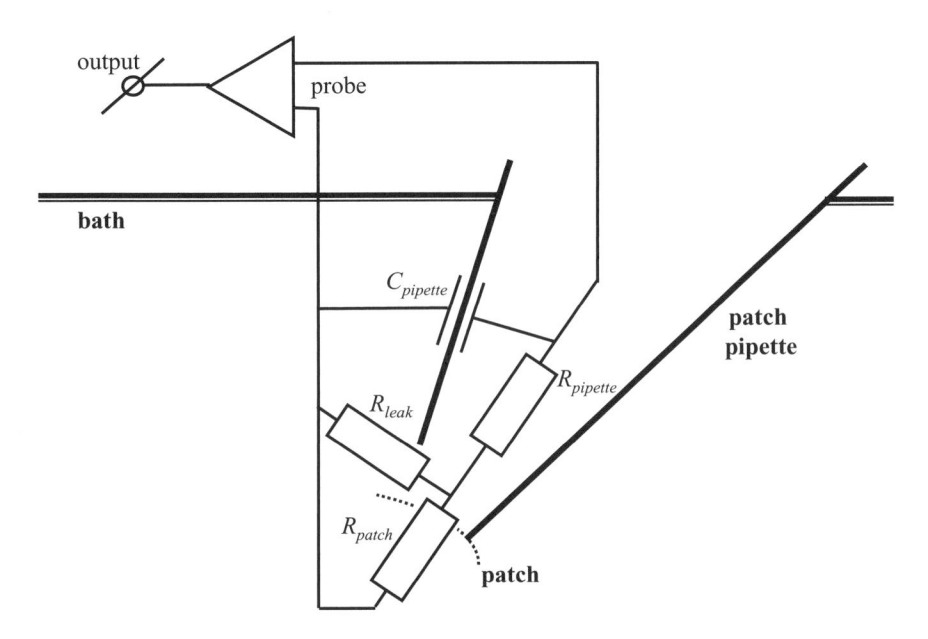

Figure 2.27 Equivalent circuit for excised patch configurations

The most important resistor, as in the cell-attached patch configuration, is the patch resistance R_{patch}. The circuit is meant to monitor current through this resistor. The parallel leak resistor R_{leak} should be very high to minimise short-circuiting of the patch current. Although the pipette resistance $R_{pipette}$ should be as low as possible, in practice the primary consideration regarding the pipette tip size is the amount of membrane that the experimenter wishes to catch (see Section 4.3). The pipette capacitance $C_{pipette}$ can be significant within the range of amplification and time resolution required for single-channel recording and must be compensated for.

In the absence of a cell, voltage clamp of an excised patch is straightforward. The holding potential is the same as the potential 'seen' by the patch under investigation, apart from a small voltage drop over the pipette resistance $R_{pipette}$. Great care, however, should be taken in applying sign conventions correctly, because these are the reverse of the two excised patch configurations and easily confused: the holding potential as indicated on a patch clamp amplifier indicates the potential of the patch electrode with reference to ground (the bath). In the case of an outside-out excised patch, where the pipette faces the cytosolic side of the membrane, this is also the patch potential. Conversely, in inside-out patches the electrodes, and therefore the potentials, are reversed (Figure 2.28). The same goes for the currents recorded from these patches.

outside-out excised patch   **inside-out excised patch**

HP = -60 mV   HP = -60 mV

E_{patch} = -60 mV E_{patch} = +60 mV

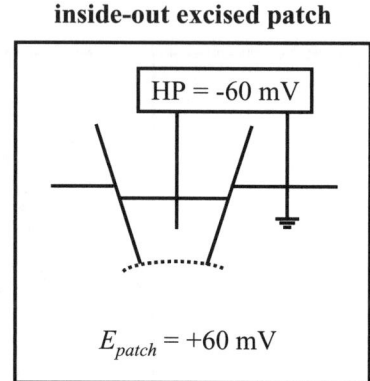

Figure 2.28 Translation of holding potentials (HP) to patch potentials in excised patch configurations

3
Requirements

Although patch clamp set-ups range from a simple rig to the most elaborate patch clamp arrangements in which a large number of variables are carefully controlled, there is a basic set of conditions that must be met in all cases for patch clamping to work. These conditions will be discussed in this Chapter. In brief, because patch clamping involves the placement of a glass micropipette onto a cell to form a tight seal, the basic elements of a set-up are

- a platform with minimal mechanical interference;
- a microscope for visualisation of the preparation;
- manipulators to position the micropipette;
- electronics to perform stimulation, recording and analysis in an electrically clean environment.

These elements will be discussed in turn below.

3.1 The Platform

3.1.1 Stability: vibrations and drift

Manipulation of electrodes and recording from preparations on a microscopic scale obviously require a very stable platform. How stable is 'very stable'? A distinction must be made here between fast movements and slow movements or *drift*. The latter is most often caused by slow changes in the micromanipulator's mechanism, and will be discussed in Section 3.2. Quick (<100 ms) movements or vibrations imply large accelerations and are often fatal to seals, even if the distance travelled is small relative to the pipette tip and cell size. Fast movements are caused by transferral of external movements to the preparation and/or pipette tip, therefore the best way to prevent vibrations is to isolate the preparation and the micropipette from possible vibration sources. Isolation starts by finding the right spot in the building for the rig.

3.1.2 Where in the building should the set-up be placed?

Although electrical considerations, such as the vicinity of a power plant, can sometimes play a role in determining the location of a rig for cellular electrophysiology, it is mostly the stability of the floor that will be the determinant. There is an enormous variety in stability or dampening characteristics between and within buildings, and sometimes even within a room. Choosing the wrong spot can definitely mean doom for your patch clamping enterprise, no matter how good your equipment is. Unfortunately there is no sound relation between stability and easily observable building traits, such as age. Here is some widely believed folklore:

- 'Placing the rig near an elevator shaft is good because that part of the building is structurally reinforced.' This might be true, but the lift itself generates vibrations and possible electrical interference. The advantage can so easily turn into a disadvantage.

- 'Placing the rig near an outside wall is good because outside walls are the skeleton of the building and therefore the most stable.' This is not always the case and depends on the design of the building. Particularly newer, multistorey buildings often have their outer walls attached to a steel frame structure that forms the real skeleton. What is perhaps nearer to the truth is that it is wiser to place a rig near a wall than in the middle of a room, where building resonances will have higher amplitudes (Figure 3.1).

- 'The closer to the ground floor the better, because this reduces the amplitude of building resonance.' This is usually true, but there are building constructions where floors are *suspended* from a steel frame, so a higher floor can be more stable than lower ones because it is closer to the supporting structure.

- 'Concrete floors are better than wood.' This is not written in stone. The advantage of a concrete floor is that it is hard and therefore

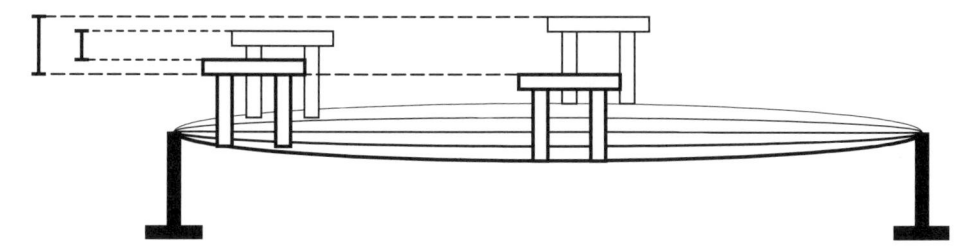

Figure 3.1 A floor area is prone to higher amplitude vibrations away from support structures

relatively resistant to low-frequency movement compared with a wooden floor. However, its stiffness makes it susceptible to the transfer of high-frequency vibrations whereas wood has much better dampening properties in this respect. What is better depends ultimately on the combination of support structure and floor material, which determines the resonance frequencies that can make the set-up unstable. The only general rule relating to floors is that more material means better stability, following the law of mass reaction: it is more difficult to move a large mass than a small one. All other things being equal, a concrete floor will therefore be better than a wooden one because it is heavier.

It is clear from the above that if you start afresh and you have the luxury of picking your spot or being able to swap space with somebody, then you might want to start by speaking to the Estates Department. Find out where the supporting structures in your building are and what materials are used. If possible, find a quiet spot where there is no busy thoroughfare or communal room nearby, keeping in mind that the most disruptive vibrations come from the transport of gas tanks and other heavy goods by trolley or forklift truck. The previous considerations can all help in finding the best spot, but it is still not an exact science and you might need to experiment. If you do, be sure to use as much of the ultimate rig (table, microscope, manipulators) as possible because the stability of your rig is dependent on a combination of the properties of all these elements.

3.1.3 Anti-vibration tables

There will nearly always be residual transfer of vibrations from the building to the rig, therefore the table on which the microscope and manipulators are placed must ideally have dampening properties. In addition, the table must be sturdy and stable when the experimenter performs manipulations on it. In some cases these requirements are met simply by making the table very heavy, with perhaps some dampening material between the base and the tabletop to absorb high-frequency vibrations. 'Do-it-yourself' enthusiasts can try a combination of the following examples that I have come across or create their own construction.

For the base:

- A U-shape from large bricks, with or without cement.
- A metal pipe frame filled with sand.
- A U-shaped piece of concrete taken from a balance table.

For dampening material:

- Pieces of rubber, e.g. bicycle inner tubes.
- Hard foam.
- Tennis balls.
- Buckets of dry sand.
- Commercially available dampers.

For the tabletop:

- A slab of concrete.
- A slab of metal, with a thickness of about an inch.

Both of these can come from discarded balance tables, forgotten in a basement room. Be sure that the floor can hold the construction you have in mind!

Gas has far superior vibration isolation properties to any solid material, so commercially available anti-vibration tables (Figure 3.2) work on the principle of air cushions in the table legs that support a very heavy tabletop. The tabletop is kept afloat by a gas source, which can be a nitrogen tank or an air pump. A feedback valve in each leg prevents over-inflation of the cushions. These tables are far better than home-made

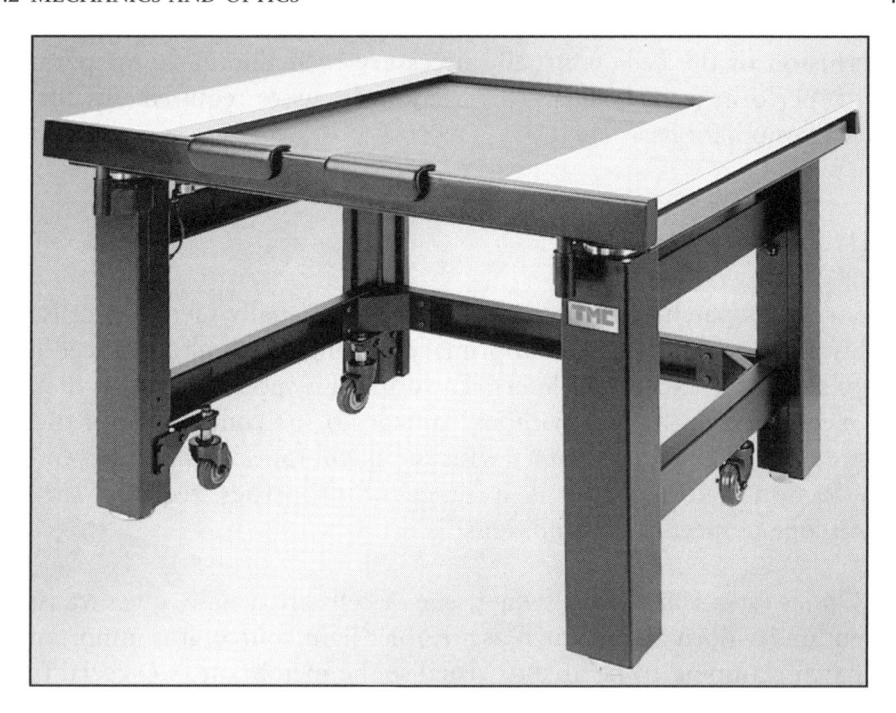

Figure 3.2 A commercially available anti-vibration table (reproduced by permission of the Technical Manufacturing Corporation)

solutions but not always necessary, whereas in cases of very unstable floors it is possible that even they cannot compensate sufficiently. However, it is likely that these 'air tables' can be used in more places in a given building than home-made constructions. This is important in the realistic situation of an experimenter having a limited choice of location when setting up. Air tables can be obtained in different sizes and with tabletops that allow constructions to be solidly attached to them.

3.2 Mechanics and Optics

The tip of a patch pipette has a pore of a few microns and cells range from about ten to hundreds of microns in size. For the pipette to form a tight and stable seal with the cell membrane, the contact must be made cleanly and precisely. A high quality microscope and micromanipulator are an absolute necessity to perform this adequately. Specific to patch clamping, it is essential that a micropipette is connected to a small pressure system to obtain a patch seal. Once the desired patch configuration is established, superfusion of the cells is usually necessary to maintain the preparation and to perform experiments. This Section discusses requirements for the above manipulations.

3.2.1 The microscope

In most cases patch clamping is performed on visually identified cells, i.e. freshly dissociated cells, cells in primary or long-term culture or cell lines (more about the exceptions later). The crux of a successful seal is that the cell membrane is reached without damage to, or contamination of, the pipette tip and that the contact with the membrane is full and even (see also Section 4.1.3). There is a range of properties required for the microscope to meet these conditions:

1. *Optics type.* Unstained, living tissue or cells are usually quite transparent under normal light such as a Köhler light source. It is important in patch clamping to be able to visualise the membrane, so special types of microscopy are usually applied. For single-cell work, these are phase-contrast and Nomarski microscopy. Detailed explanations of these techniques are outside the remit of this book, but in principle Zernikeian or phase-contrast microscopy exploits differences in the refractive index for different structures (e.g. the membrane and the cytoplasm) to elicit brightness differences between them. This occurs through interference of normal light transmitted through the object with light of which the phase was rotated before it reached the object. The resulting pattern of extinction is dependent on the refractive index of the object. In Nomarski or interference contrast microscopy, two beams of polarised light are used to detect differences in distance from the objective, resulting in a very useful three-dimensional type of effect. However, Nomarski optics are considerably more complicated and expensive than phase-contrast optics. Patch clamping of visually identifiable cells in brain slices is particularly challenging, and a combination of Nomarski optics and low-scattering infrared light is sometimes used here. This obviously requires a sophisticated video system.

2. *Working distance.* With a 'normal' microscope where the objective approaches the preparation from above, the space left between preparation and objective is too small to position a patch pipette. The magnification needed to visualise both the cell and the pipette tip

3.2 MECHANICS AND OPTICS

rarely allows the use of objectives of less than a magnification of 40, which need to be very close (less than a few millimetres) to the preparation. Patch clamp rigs for single-cell work therefore require *inverted* microscopes where the objectives are placed under the preparation (Figure 3.3). There are two measures that can be taken to allow the objective to come close enough. The cells can be grown on thin substrates such as coverslips that can be mounted directly in the microscope, instead of on petri dishes, or, if this is impossible, special long-working-distance objectives can be obtained but these are very expensive.

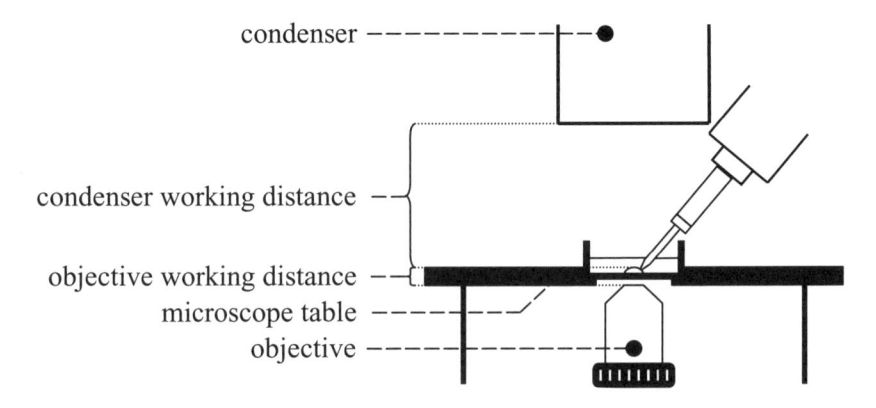

Figure 3.3 Working distances on inverted microscopes. Note that in microscope specifications these distances are given to the focal point and are therefore larger than the actual manoeuvring room

If the objective is situated below the preparation, then the light source above it determines the room available for the patch pipette, tubing, etc. In the case of standard illumination this usually does not pose a problem because the condenser can be placed in a high position. Unfortunately the type of optics used most widely in patch clamping, i.e. phase-contract microscopy (see previous paragraph), requires the illumination unit to be set at a specific and not too large a height above the preparation. In practice this height usually determines the working distance, and it is worth comparing manufacturers' specifications and letting this parameter be a consideration in the choice of microscope. Why is it so important? The working distance determines the range of angles of approach of the pipette. The angle should be as perpendicular as possible if cells are flat or in monolayer, or in most cases if brain slices are used. Conversely, the approach should be as horizontal as possible in the case of cells protruding sideways from explants or tissue slices. A large working distance provides the greatest flexibility. In addition, it must be remembered that replacing pipettes is done very often and facilitating this by having a large working distance becomes an important comfort factor.

3. *The microscope platform.* The platform or table should be both stable and easy to adjust. Modern microscopes have large tables that are adjustable in the x- and y-directions, sometimes by a flexible boom. They are even large enough to support perfusion systems, electrodes, etc. in addition to the preparation itself. The situation is much trickier with smaller microscope tables intended for studying microscope slides. Here, adjustments often also lead to unwanted displacements in the z-(vertical) direction, especially if the table is loaded with perfusion tubing (see Section 3.2.4). Smaller microscopes cannot usually support manipulators, thereby increasing the demands on the stability of the platform (see Section 3.2.2).

4. *Flood management.* It is virtually inevitable that the bath will flood at times. The microscope, particularly around the objective's revolver, should be waterproof or easy to disassemble and clean. Some microscopes require a qualified maintenance engineer to attend to this!

5. *Power supply.* The lamp power supply is usually a very 'dirty' alternating current, potentially creating rampant 50–60 Hz noise within your carefully crafted Faraday cage (more detail on noise in Section 3.4.1 and elsewhere). Microscopes vary enormously in the level of shielding of this supply and it seems that successful designs are due to serendipity, because I have never seen a manufacturer use this point as a design feature. The power supply usually contains a transformer that can be built into the lamp housing on top, in the base of the microscope or in an external unit. The latter two designs have the possible disadvantage that there must be a lead running from the transformer to the lamp housing, and this lead, if unshielded, is radiating noise. It is possible to circumvent lamp noise by using a stabilised DC power supply, but this can dramatically shorten the lamp life if not used exactly right. A simple solution can be to apply aluminium foil connected to ground to cover the transformer unit and/or the lead. Be careful with wrapping the lamp housing, because lack of cooling might shorten lamp life or even start a fire.

3.2 MECHANICS AND OPTICS

6. *Extra ports*. It is often desirable to have camera ports on the microscope. A video camera allows monitoring of the configuration. This has the following advantages:

   - The visual signal can be observed simultaneously with the electrical signal (on oscilloscope or computer screen).

   - The experimenter does not risk mechanical interference by accidentally touching the microscope with his or her head.

   - If a large Faraday cage (Section 3.4.1) is used, then this can be closed before the patch is made, eliminating the risk of breaking the patch later.

   - The experiment can be recorded. For example, smooth-muscle cells can be stimulated to contract during a patch clamp experiment. The contraction can be quantified *off-line* by video analysis.

   - Stills from a video camera or a photo camera can be used to document the cell investigated and identify the cell in case the preparation is fixed for further histological study.

   - Video footage is very useful in teaching and training.

Camera ports together with add-on equipment can also open up possibilities for immunofluorescence and calcium fluorescence. This can be an important consideration when a new microscope is purchased, even when the fluorescence application is not immediate.

Patch clamping in brain slices

In its simplest form, patch clamping in slices is a semi-blind technique, i.e. the precise cell being patched is not visualised during the experiment. The microscope here can be a high quality preparation microscope. Areas in the slice that contain many cell bodies (usually targeted for patching) are more transparent than areas with white matter, and can be recognised as dark areas in reflective light, or as light areas if the light comes through the slice. For example, in a hippocampal slice the cell bodies of CA1 neurones can be seen easily as a bright line using transillumination. The working distance with a preparation microscope is quite generous, even though the objective is on the 'business' side of the preparation. If the microscope comes with its own light, then this is likely to be insufficient and is usually replaced by a cold light, of which the lamp housing/power supply unit can be placed outside the Faraday cage (see Section 3.4.1).

As mentioned earlier, patching visualised cells in slices is possible but can become costly if a combination of Nomarski microscopy and an infrared video system is required. Alternatively, in some slice types cell bodies can bulge out from the sides, where they can be seen under a normal (non-inverted) microscope. The consequently very small working distance is one of the additional challenges of this type of recording over blind patching.

3.2.2. Micromanipulators

The main function of a micromanipulator in a patch clamp set-up is to position the micropipette tip onto the cell membrane in a controlled manner. Additional micromanipulators are sometimes used, depending on the experiment. For example, electrical field stimulation can occur through wires or plates that are positioned by a second manipulator, or local superfusion can be performed using micropipettes in combination with a 'puffer' system (see next Section). In positioning a micropipette onto a cell, the scale of manipulation quickly changes over orders of magnitude, from roughly placing the pipette over the preparation (10^{-2}–10^{-3} m), to finding the perfect spot on the cell membrane (10^{-5}–10^{-6} m). For this reason, most manipulators have three scales of manipulation:

- Adjustable positioning of most of the manipulator around free axes of rotation (which can be fixed by tightening bolts) is used for replacing micropipettes and positioning the pipette over the preparation.

- A three-axis coarse mechanical manipulator with a range of 1–3 cm is used to lower the pipette into the bathing solution and to approach the cell to within about 100 µm, or a range where the pipette and the cell can be seen in the same field.

- A fine mechanical or hydraulic manipulator is used to position the pipette onto the cell.

Examples of common manipulators are presented in Figure 3.4. In some mechanical manipulators there is no course manipulation but the range of

Figure 3.4 Manipulator types. (Left and middle) A coarse mechanical manipulator combined with a hydraulic, remotely controlled micromanipulator (reproduced with permission of Narishige Scientific Instruments Laboratories). (Right) An all-mechanical manipulator (reproduced with permission of Märzhäuser Wetlar)

fine manipulation is sufficiently large. In addition, in these manipulators one or more of the axes can be motorised, so that movement along the axis is driven by a small, remotely controlled electromotor. There are also manipulators driven by a piezo element, which are particularly suitable for quick and precise movements required for intracellular recording (see Section 2.3.2).

Mounting the manipulator

Three major aspects should be considered in placement of the manipulator. These are:

1. *Minimising vibration.* Manipulator units are prone to oscillation in response to vibration, because the probe and micropipette (and sometimes parts of the manipulator itself) form moments that can resonate and transmit vibrations towards the tip, with disastrous consequences. There are a number of measures that can be taken to minimise this:

    - Reducing vibration amplitude by increasing mass. For example, mechanical manipulators such as shown in Figure 3.4 (right) are of a very robust, heavy construction.

    - Reducing moments. Any extension of the manipulator unit increases oscillation amplitude at the pipette tip. The probe should be mounted on the manipulator as centrally as possible. The three different axes of manipulation, both coarse and fine, should be built up in such a way that the whole unit is as small as possible. Note that this strategy can be thwarted by a robust construction because of its size. For this reason hydraulic manipulators are light but also compact and can be mounted straight onto the microscope stage.

- Another advantage of mounting the manipulator directly on the microscope is that there is a certain degree of cancellation of vibration because they move in unison. Unfortunately this is not possible with small microscopes or heavy mechanical manipulators.

- Avoiding weak spots. Any manipulation system is only as stable as its weakest link. Notorious weak spots are lightweight attachments (single bolts, hinges) within the manipulator or on a microscope table, feeble microscope tables, or loose tubing or cables. The latter not only create moments but can transmit vibrations from outside the anti-vibration table. This can even be a cause of problems using hydraulic manipulators where the control unit (Figure 3.5, centre photograph) is away from the stabilised area and connected to the manipulator by means of poorly secured tubing.

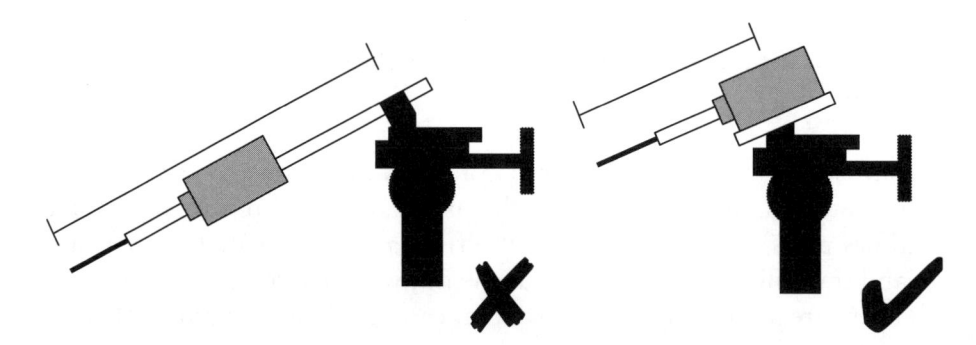

Figure 3.5 Moment reduction in a manipulator unit

2. *Optimising practicality.* No matter what configuration of manipulator is chosen, it will always take a while and usually a number of micropipettes for the experimenter to get used to the system. Although it is difficult to make specific recommendations because of the different

3.2 MECHANICS AND OPTICS

types of manipulator available, much frustration can be prevented by ensuring the following:

- It must be easy to replace a pipette. This will be an often-repeated procedure and should be possible without disruption and with the least amount of manipulation possible. Be aware of nuts and bolts coming loose with repeated movement of components.

- If possible, the coarse manipulator should be in line with the horizontal (i.e. not tilted). Positioning of the pipette over the preparation is easiest when the axis of view (perpendicular to the horizontal plane) lines up with the manipulator z-axis. The same argument goes for the fine manipulator, but here another factor comes into play – the pipette is likely to make the best contact with the cell if the axis of movement coincides with the axis of the pipette. This axis will never be perpendicular to the horizontal, so movement will be simultaneously in the x- and z- directions. As a result, a great deal of care and judgement is required to position the pipette tip on the desired spot on the cell. This is one of the sporting aspects of patch clamping!

- There are left- and right-handed versions of mechanical manipulators and hydraulic remote control units. The controls on them will be facing towards the experimenter and rotate in a logical way if positioned correctly. Although eventually you will get used to inverted manipulators, it will save time and reduce frustration to avoid this, particularly if several people use a rig or if more than one type of manipulator is used.

3. *Minimising drift*. Drift is the slow, continuous movement of the pipette in relation to the preparation. It is caused by anything that is out of static equilibrium and has the opportunity to exert its force onto an element that can be displaced. The generality of this statement is reflected in the many possible causes of drift. However, close examination of the definition is helpful to reduce drift: eliminate the sources of force and the possible movement of elements.

   What are the sources of force? Any weight bearing down on an element that is not securely fixed, like a weak link in the manipulator, will cause drift. Cables and tubing will have a spring action and can easily cause drift because they often attach very near the pipette. (An important force that is often overlooked in this context is *torque* in cables and tubing.) Temperature and humidity affect materials differently and therefore any changes in these variables will cause displacement forces. Some types of drift eventually eliminate themselves: a newly constructed rig will nearly always show drift but will settle down over time. The more persistent forms are caused by repeated disruption of the equilibrium, e.g. changing the pipette or switching on a superfusion or pressure system. The elements in a manipulation system that are most commonly moved are the pipette holder (exposed to torque, spring and pressure forces from tubing) and the pipette itself (not tightly secured in the holder). The turning on or off of a superfusion system can also cause the preparation to move, by vibration *and* drift!

Modern mechanical and hydraulic manipulators (particularly the water-filled as opposed to the older oil-filled ones) show very little drift if used properly. This includes not overloading them, and tightening the controls appropriately. Hence, the amount of drift seen under practical patch clamp conditions bears little relation to the manipulator manufacturer's specification. Generally, reduction of drift is achieved by robust construction (to reduce possible movement of elements) and stabilisation of conditions. If there is still persistent unacceptable drift, it is useful to review the mechanical manipulation during the experiment and try to identify the source of the drift force and the moving element.

3.2.3 Pipette pressure

When the pipette approaches the cell, positive pressure on the pipette fluid is required to keep the tip from contamination by debris in the bath. Conversely, in the formation of a seal some delicate suction needs to be applied. For these reasons patch pipette holders possess an auxiliary channel connected to a tubing system. The system must be able to provide constant pressure of about 5 cmHg maximum and a switch to a manual – or rather *buccal* – element that allows suction by mouth. The pressure device is usually a U-tube filled with water or mercury. In a well-sealed system it is possible to apply pressure by mouth and then close the system, but compared with a U-tube the disadvantages are that the pressure is difficult to quantify and leaks are not detected immediately. It is important that the tubing is of a stiff variety so that suction is well controlled. However, stiff tubing increases the build-up of spring action and/or torque and so contributes to drift (see previous Section). These problems can be

3.2 MECHANICS AND OPTICS

minimised by firm attachment of the tubing to the manipulator assembly. The whole system is home-made but plastic three-way blood transfusion taps and 1 ml syringes (without the plunger) as replaceable mouthpieces are very effective. A diagram of a typical system is presented in Figure 3.6.

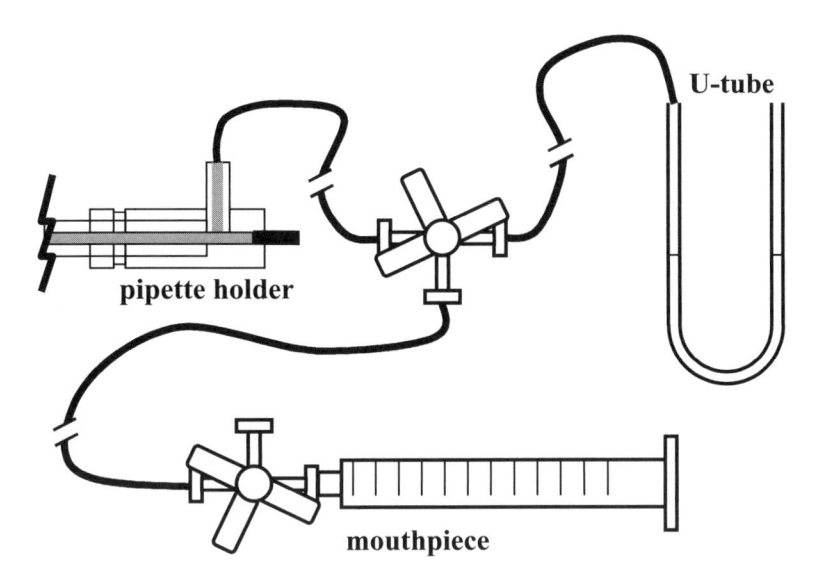

Figure 3.6 A simple but effective pipette pressure system. All elements except the pipette holder are located outside the anti-vibration area

3.2.4 Baths and superfusion systems

In most cases it is a requirement that the bath in which the preparation is immersed is superfused. Both the bath and the superfusion system are hugely variable depending on the preparation, the microscope platform and any special requirements. The following is intended to introduce some practical, low-cost examples with their advantages and disadvantages.

Baths

The basic requirements for a preparation bath are that the preparation can be visualised, manipulated and superfused, that there is appropriate access for the micropipette and that a ground electrode can be accommodated. Often some sort of temperature control is also necessary. In its simplest form, the bath is a petri dish (10 cm^2 or smaller) with a supply of superfusate on one end and a drain on the opposite side, and a silver wire or pellet dipped into it from the edge. The supply, drain and ground electrode can be held in place by little magnets (if the preparation holder is metal) or plasticine. However, the major problem with this is that superfusion of a circular bath is very uneven. If technical support is available, then perspex baths can be made to the experimenter's specifications (Figure 3.7). Some possibly useful features are listed below:

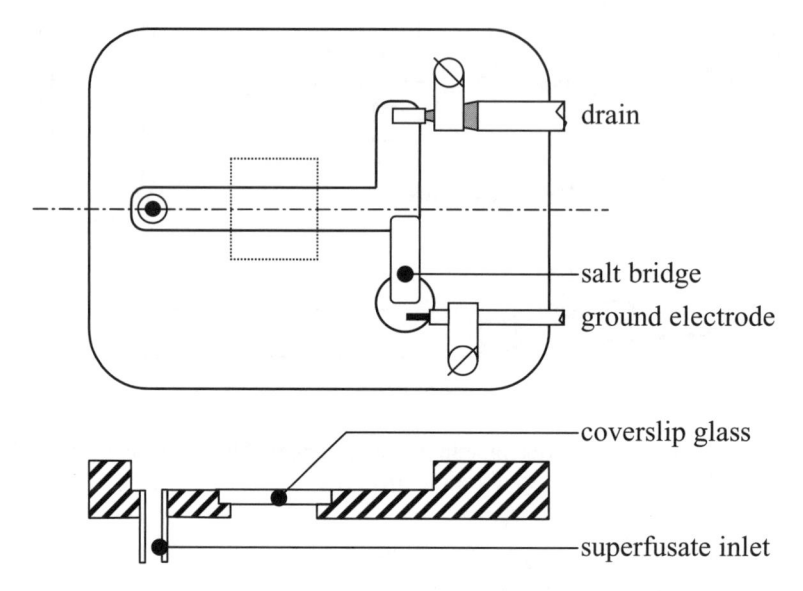

Figure 3.7 Example of a design for a perspex bath

- To obtain stable superfusion, the shape of the bath is elongated so that a near-laminar flow is obtained. This can be tested by perfusion of dye solution. For various reasons a number of extra channels are sometimes constructed but this will hamper perfusion, so it pays to keep it as simple as possible.

- The superfusate inlet can be on the bottom of the well, whereas the drain is usually a syringe needle (see next Section).

- Both the drain and the ground electrode can be fixed to the bath but it must be remembered that they need regular servicing, so the fixing should be easily reversible.

- A ground electrode can be placed directly into the bath or into a separate well connected to the bath by a salt bridge (see Section 3.3.2 for details).

- A water jacket can be built into the bath construction to maintain a chosen temperature. However, water flow can cause vibrations and the construction, with its afferent and efferent water tubing, can be cumbersome and unwieldy. Temperature control is performed more easily by preheating of the superfusate, although this requires continuous superfusion if the temperature is to be kept constant.

The shape of the bath is restricted by the substrate of the preparation, but the substrate is often flexible, e.g., coverslips can be cut to size, and optimising superfusion is well worth the extra effort. Big adjustable microscope platforms usually have threaded holes so that a perspex bath can be mounted easily with bolts. The use of plasticine or tape is a bad idea (even compared with relying on gravity) because it can lead to drift.

Finally, baths are also commercially available. This can be an easy solution if you do not have technical support but have money to spend, or if you have special requirements. For example, baths can be obtained with a good heating system combined with a way of controlling the atmosphere above the bathing fluid. Such a bath would be very difficult to produce in-house.

Superfusion systems

A stable superfusion system consists of two parts. Firstly, a set of containers with some means of selecting one of them, connected to the bath inlet by tubing through which the flow rate can be adjusted. This can also include a heating device for temperature control. Secondly, drainage with an adjustable flow rate to match the inflow is required.

By far the best way to ensure a constant flow is to use gravity as the driving force for both inlet and drain. The next best thing, a peristaltic pump, will cause oscillations in the bath fluid level, particularly if used for the inflow. As an example of a simple superfusion system, I will describe one that I have used in the past but also seen in many places in various guises, so the origin is unclear. The containers' unit is made from a large bottle with a drain at the bottom for the basic extracellular solution and several 60 ml syringes (without the plungers) for extracellular solution with drugs added. These are connected by a series of blood transfusion taps. The final tap in the sequence holds a (complete) 10 ml syringe to aid initiation of perfusion and cleaning. The whole assembly is mounted onto a tripod stand (Figure 3.8). The height of the containers in relation to the bath determines the flow rate. The optimum flow rate is a compromise between a fast flow, which promotes fast response times (short delays between switching solutions) and steep concentration gradients, and a slow flow, which saves drugs and causes little mechanical disruption. The usual range is 2–10 ml min^{-1}.

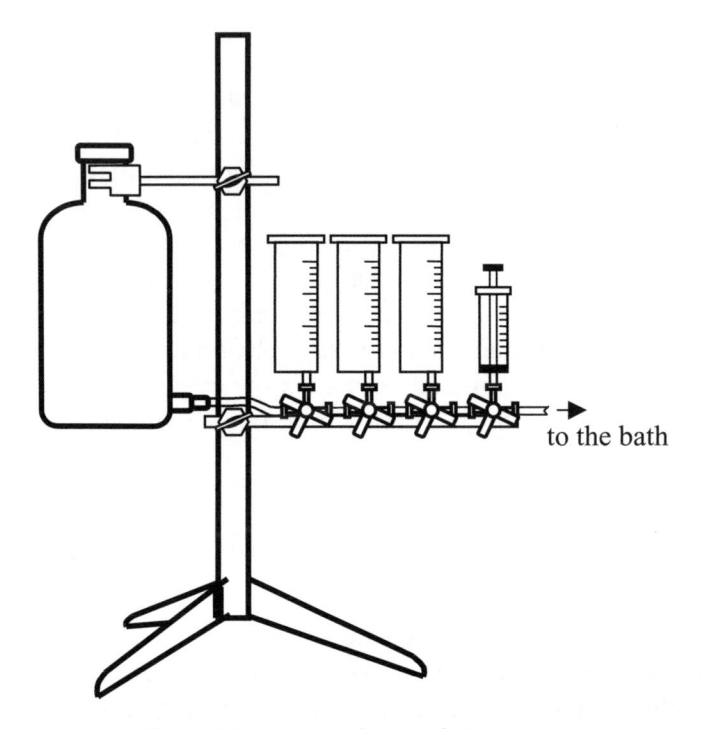

Figure 3.8 A versatile superfusion source

Of course any of the elements can be replaced by something similar that happens to be around. Although laboratory scavenging is a deplorable practice, it must be said that most elements will probably be available in most laboratories if sought by the resourceful.

This simple system can be expanded in two important ways:

1. If gas dissolved in the superfusate requires control, e.g. if the solution is buffered by carbonate and the solution must be saturated with

oxygen, then the containers can be connected by bungs to (infusion-) bags filled with the required gas. The bags can be of any type, as long as they are not elastic.

2. A temperature control device can be built into the system to preheat the superfusate before it enters the bath. The device can be electrical (Peltier element) or a water jacket. The former has the disadvantage that on–off switching of the heating coil can cause inductive noise, whereas the latter requires a lot of extra tubing and can increase the delay between switching solutions considerably. However, a water jacket can also be used as a cooling device. It is possible in principle, to set up a feedback system between a temperature sensor (which can be very small and placed close to the preparation) and the heating device (Peltier power supply or heater for the water jacket) to obtain automatic temperature control.

The materials used in the system should be considered. In most cases silicone tubing will be the most inert material, e.g. it is less sticky to lipophilic compounds than rubber-based tubing. Lubricants in glass syringes can cause the same problem. It is important that metal connectors are avoided because there is the possibility of contamination of the solution by metals.

The drain controls the fluid level in the bath. Because the drainage rate must be equal to (which is not practical) or higher than the inflow rate, the drainage system will, in practice, draw in air bubbles. The size of the bubbles is proportional to the size of the fluid level oscillation (see Figure 3.9), which should be kept to a minimum.

If the bubble/fluid oscillation becomes very big, then there is the chance that the tubing will empty in one swoop, thereby removing the suction, and a flood will result. Bubble size is determined by a combination of drain shape, tubing qualities and suction. Unfortunately, a successful combination cannot be prescribed and has to be found by experiment in each new situation. This can test the experimenter's patience and, although gravity is the best source of suction, peristaltic pumps or aquarium pumps can be used to simplify the equation. The shape of the drain is critical and an oblique shape such as the tip of a syringe needle is more conducive to the generation of small bubbles than a straight one. A metal syringe needle for a drain might be less of a problem regarding metal contamination than metal parts in the inlet system, because the drain only moves solution *away* from the preparation. However, obliquely-cut fine tubing can also be used. If using gravity, the tubing type and bore need to be selected by

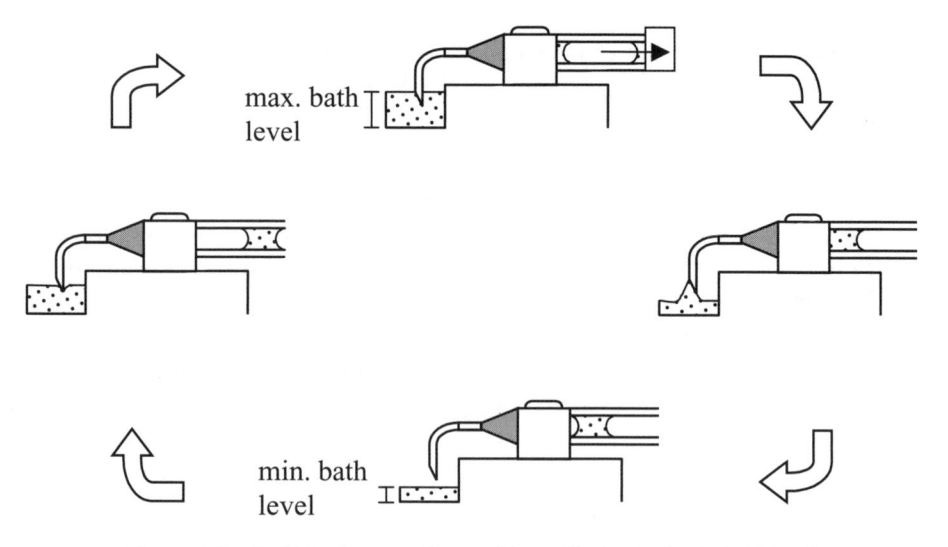

Figure 3.9 Bubble size correlates with oscillation in bath fluid level

experiment, but experience shows that the chances of success are greatly improved if the tubing is dirty on the inside! This is probably due to a higher surface friction that slows down flow, thereby postponing complete emptying of the tubing and suction failure. The suction generated is determined by the (vertical) tubing length and the amount of fluid in the tubing. Drainage is initiated by a pull-through tap and syringe at the lower end of the tubing.

Special perfusion applications

Three special systems are briefly mentioned here: local superfusion, electrophoresis and micropipette perfusion. Local superfusion can provide very quick response times and exposure of only part of the preparation to drugs, allowing more than one cell in a preparation to be treated as naïve. It is also extremely economical on drugs because very low volumes are involved. It is usually performed using a 'puffer' system consisting of a set of micro-barrels connected to a pressure system. Superfusion is controlled by regulating the pressure and/or moving the barrels to expose the cell to different flows (Figure 3.10).

3.2 MECHANICS AND OPTICS

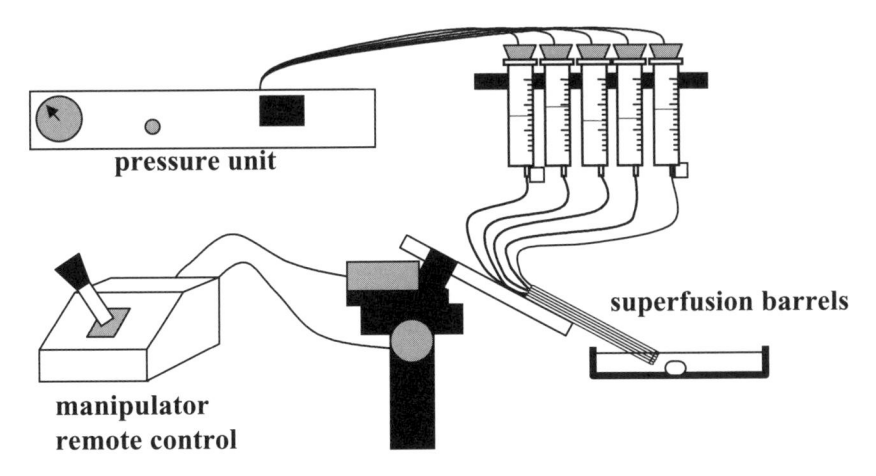

Figure 3.10 Principle of a local superfusion system

The drawbacks of local perfusion are that a complicated device is required and it is not easy or always possible to calibrate the system, i.e. it is difficult to know the concentrations of drug that are reaching the cell.

The same problem applies also to the technique of electrophoresis. Not strictly a superfusion system, the method utilises the charge properties of the drugs that are to be administered. A potential difference is set up between a pipette containing the drug and the bath. The polarity, strength and duration of the potential will determine the direction of movement of charged molecules within the pipette and the amount applied. Although exact effective concentrations are sometimes difficult to ascertain, the method is very reproducible owing to the precise control of potential.

Micropipette perfusion is a very powerful technique that allows replacement of the pipette solution while an experiment is in progress. It circumvents the problem that in the whole-cell mode the intracellular solution cannot be changed. Normally the effects of cytosolic factors can be tested only by comparing data obtained from groups of cells using different intracellular solutions. In micropipette perfusion experiments each cell can be its own control. The principle is very simple (Figure 3.11): an extra channel is introduced through the pressure tubing and connected to reservoirs with different intracellular solutions. In the whole-cell configuration suction and pressure are applied to change the pipette solution and, subsequently (with some delay), the intracellular solution. However, in practice the technique is very challenging. Obstacles include the complicated construction, the fact that the pressure and the suction must balance perfectly, the mechanical problems of manipulation and that the whole

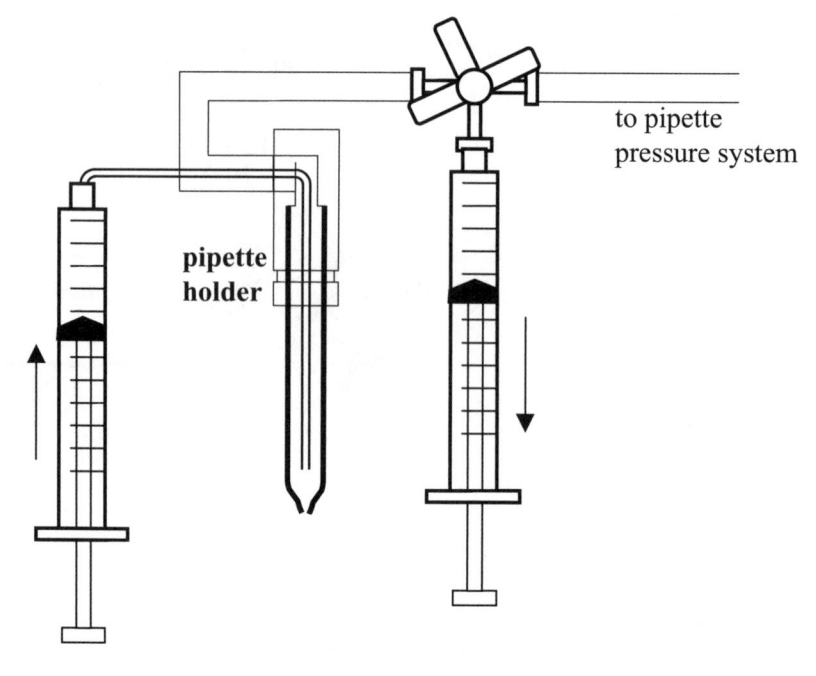

Figure 3.11 Principle of a pipette perfusion system

perfusion system is effectively connected to the very sensitive recording electrode, creating a big noise problem (see Velumian *et al.*, 1993, for more details).

3.3 Electrodes and Micropipettes

Confusingly, the term 'electrode' is used in the literature for related but different things. In this book an electrode is always a conductive solid that makes electrical contact with a fluid. However, often the term is used to designate a complete electrode/micropipette/pipette-holder assembly. In this case it is better to speak of a microelectrode to avoid confusion. In patch clamping usually only two electrodes are used: the pipette electrode and the bath electrode. In special cases field stimulation electrodes can be added, although they are not directly associated with the patch clamp technique. In this Section, essential properties and practical implementations of electrodes will be discussed.

3.3 ELECTRODES AND MICROPIPETTES

3.3.1 Solid–liquid junction potentials and polarisation

Redox reactions (electron transfer) between metals and salt solutions will cause the establishment of a potential between the two media. In an experimental situation, if the two junction potentials are not equal, they can create an artefact in the form of an offset current that is wrongly cancelled by the experimenter (see Figure 3.12). It is therefore important to use electrodes that are as identical as possible, both in type of material and surface area, and to use electrodes with small redox potentials, such as platinum or silver chloride. In this way, identical but opposite solid–liquid junction potentials on either side of the preparation in the circuit will cancel each other out. Solid–liquid junction potentials should not be confused with liquid–liquid junction potentials, which are discussed in Section 3.3.4.

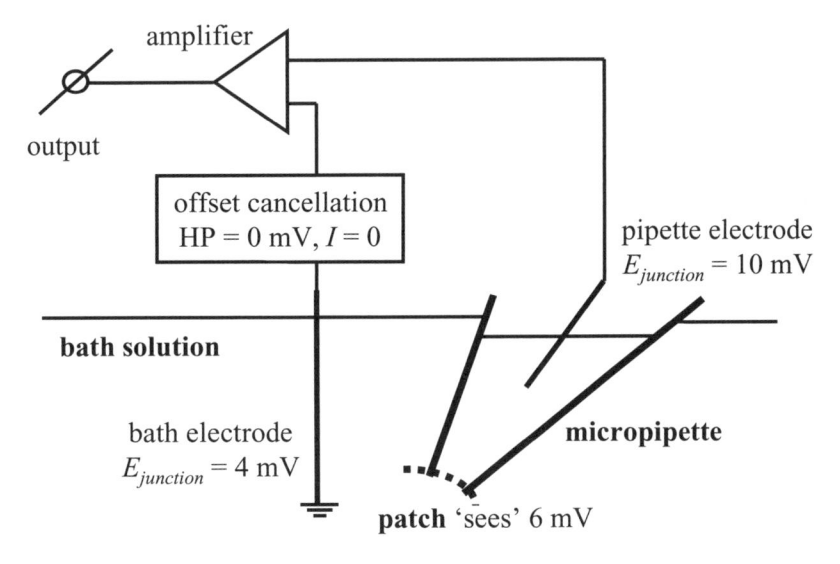

Figure 3.12 Non-symmetrical electrodes create voltage artefacts

In addition to solid–liquid junction potentials, polarisation of electrodes can occur. Polarisation is caused by the accumulation of ions near electrodes that have a DC potential difference between them (Figure 3.13). Cations move towards the negative electrode (anode) and anions move

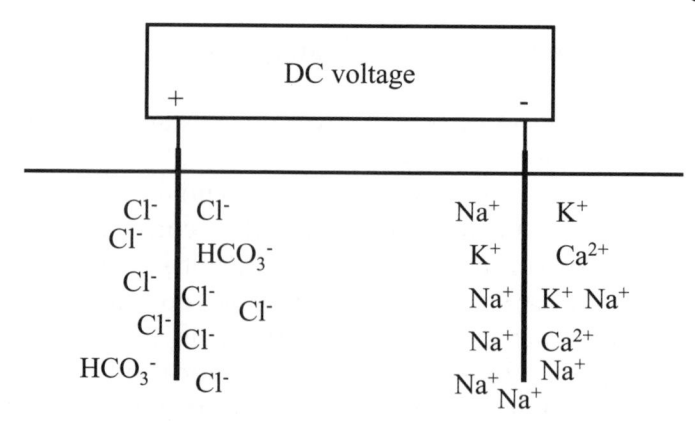

Figure 3.13 Electrode polarisation delays potential changes and increases resistance

towards the positive electrode (cathode). The accumulation has a similar effect to that seen in capacitors: changes in potential take effect with some delay. In addition, current flow is impeded. Electrode polarisation is dependent on material and shape, whereby platinum and silver chloride have the best (weakest) polarisation properties.

Platinum is expensive and difficult to solder, but silver chloride must be maintained. Silver chloride electrodes consist of silver wire or pellets that have been 'chlorided', forming an AgCl layer on the outside of the wire or pellet. The chloride in solution and on the silver is in an easily reversible equilibrium, which accounts for the low junction potential property of silver chloride. The layer will wear off, particularly from the pipette electrode where repeated pipette changes will scrape away the silver chloride from a wire. (There are also pipette holders that have a pellet instead of a wire so that scraping is minimised.) One of the signs of a deteriorating silver chloride electrode is the creation of asymmetry between the electrode junction potentials, which manifests itself as an increasing offset. In addition, polarisation becomes apparent in the shape of sluggish responses, and the baseline becomes unstable. There are two common methods for chloriding silver wire or pellets to form AgCl electrodes:

1. Electrophoresis using any DC power source and a weak (0.1 M) hydrochloric acid solution. Chloride will bind to the cathode (+ pole) and hydrogen gas will form at the anode (− pole).

2. Immersion into neat household bleach.

3.3 ELECTRODES AND MICROPIPETTES

The result should be that shiny silver is coated with a dull grey layer. The former method is used most widely but the latter is much simpler. I have not detected clear differences between the results, either in polarisation properties or rates of deterioration. Both methods also clean the electrodes: in the case of electrophoresis, this is done by reversing the polarity of the power source so that hydrogen will form at the electrode, this loosens up dirt, which then can be wiped off, it is good practice to do this before every chloridation.

3.3.2 The bath electrode

In many cases a silver chloride wire or pellet immersed in the bathing solution will suffice as ground electrode. However, if the bath is perfused during the experiment there is the risk that the solid–liquid junction potential will change, and create an offset artefact. This is most likely to happen with large changes in ionic composition of the superfusate. The problem can be avoided by using a separate well for the ground electrode. This well is then connected to the bath by a 'salt bridge' (see also Figure 3.7). Such a system has the advantage that the solid–liquid junction potential is constant, independent of the superfusate. A salt bridge is usually made from rigid plastic or glass in the shape of a U-tube, filled with agar gel containing salts for conductivity. The extra well can be filled with the same solution as the pipette solution (depending on patch clamp configuration) to ensure electrode symmetry. A type that does not require an extra well consists of a silver chloride pellet at the end of a small bit of tubing. The tubing is filled with pipette solution and then dipped into an agar plate for sealing (Figure 3.14).

A seemingly trivial but important point is that care should be taken so that only the silver chloride part of a bathing electrode makes contact with the bath. Solder connections and copper wire create the same artefacts as badly chlorided silver.

3.3.3 Micropipettes

Glass micropipettes filled with salt solution are used to make the patch electrode contact area with the outside world microscopically small. This area is determined by the pipette tip diameter. The typical configuration of the pipette holder, glass pipette and pressure system was introduced in Figure 3.11. The pipette is typically made from a rod of glass with an outer

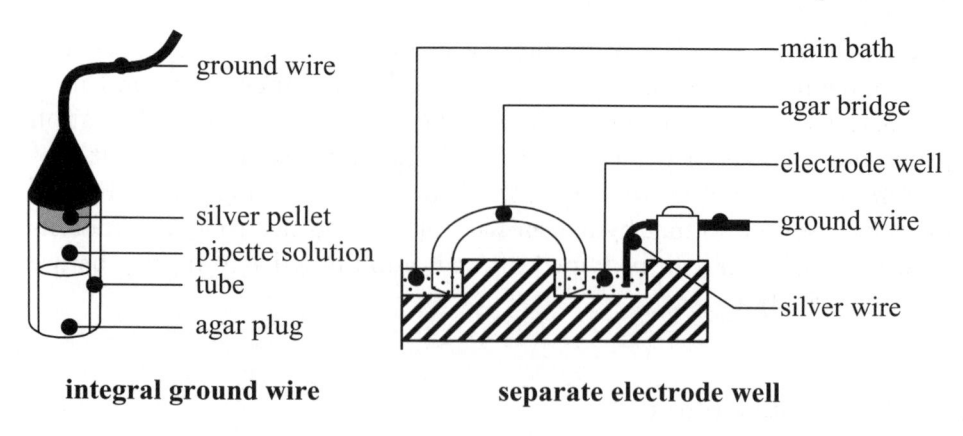

Figure 3.14 Salt bridge types for ground electrodes

diameter of 1.2–1.5 mm and a length of 10–15 cm. The tip is formed by locally melting and pulling of the rod in a *micropipette puller* (discussed in the next paragraph). The glass rods used are commercially available in different materials and sizes. The following should be considered in the choice:

1. *Outer diameter.* This needs to be considered when purchasing a patch clamp amplifier because these come with pipette holders that will take a certain diameter of pipette. In general, larger diameters (1.5 mm) are easier to work with, because:

   - when the holder contains a silver chloride wire, scraping of the wire while replacing pipettes is much less;

   - the chance of overflowing the pipette when mounting it in the holder is reduced;

   - the visually challenged among us will find it much easier to mount the pipette into the holder.

   Only if there is a special reason should the smaller diameter (1–1.2 mm) be considered. For example:

   - the set-up could impose a space restriction that favours thin diameters;

3.3 ELECTRODES AND MICROPIPETTES

- in slice patching when the preparation is viewed from the same side as the pipette approach, a smaller pipette diameter would increase the view (see Section 3.2.1);

- a large diameter pipette will have a larger surface area and therefore form a larger capacitance. In very quick recording of small (single channel) signals this can be an undesirable factor.

2. *Inner diameter.* Electrode glass usually comes in 'thick-walled' or 'thin-walled' varieties. It is easier to produce very small pipette tips as used in intracellular recording with thick-walled glass, because the amount of melted glass available allows good stretching and narrowing of the tip. In contrast, patch electrodes must be relatively blunt and therefore thin-walled glass works better.

3. *Filament.* Most electrode glass contains a filament within the rod lumen. The filament greatly aids filling of the microelectrode by increasing capillary action, which lets the pipette solution fill the tip and avoids bubbles. In principle there is no reason to buy glass without a filament (unless you are a masochist). Some workers feel that filament-less glass is better in perforated patch clamp (Section 4.2.2), because here it is required that the pipette tip is filled with a different solution to that of the rest of the pipette, and this separation might be disturbed by the filament. Although I have used filament glass successfully in perforated patch clamp, it is possible that the success rate might have been higher using the filament-less variety. There will also be no bubbles in the tip because of the suction filling, so the need for the filament is less.

4. *Rod length.* The minimum rod length is dependent on the pipette puller and is usually about 10 cm. Longer pipettes increase susceptibility to radiation noise but if the set-up is shielded well enough (see Section 3.4) then the additional length can provide a welcome increased ease of handling the probe in a tight space.

5. *Glass type.* In addition to the normal borosilicate glass (or aluminosilicate for fine pipettes) there is also quartz available. It is very expensive and requires a special pipette puller. So why use it? Quartz glass has considerably better (more neutral) light breaking and fluorescence properties than borosilicate glass and is therefore required if patch clamping is combined with fluorescence techniques,

such as immunofluorescence of intracellular calcium ions. It also has better electrical properties and is mechanically stronger than normal glass.

The glass for micropipettes usually comes in re-sealable containers. It must be kept very clean to obtain consistent pulling results and to prevent clogging of the tip. In particular, glass should never be handled where the heat is going to be applied during pulling. If the glass has become dirty, then it can be cleaned before pulling by bathing the rods in a 95 per cent ethanol solution and carefully drying them in a flame. If the glass is dusty then it also needs to be rinsed with distilled water.

Pipette pullers

Pipette pullers are often wrongly referred to as electrode pullers. They are all based on the principle that a glass rod softened by heat is pulled apart in stages until only a very thin connection exists between the two halves. The connection is then broken by a final hard pull. The combination of mechanics and electronics required for precise control has always been difficult to obtain, particularly before commercial pullers became available and pullers had to be made in-house. The problem was confounded by the fact that before patch clamping was developed microelectrodes were used for intracellular recording, which required even finer micropipettes. Now commercial pullers can provide very consistent results, although it can still be a challenge to find the proper settings and the machines can be easily disturbed if used improperly. (As a result, workers are known to protect pipette pullers ferociously, adding to the reputation of electrophysiologists being very territorial.)

In practice, a glass rod is mounted in two clamps, of which one or both can move by means of a carriage. A dual-carriage puller is usually configured so that a pull results in two identical pipettes. A single-carriage puller is simpler and therefore cheaper, but will always produce non-symmetrical pipettes, of which one often has to be discarded. The pulling force is provided by a large solenoid. The two rod clamps are located on either side of a heat source in the form of an electrically heated coil or filament, or – in more recent, expensive pullers – a laser. For extra regulation there can be a valve-controlled gas flow at the heat source to provide quick cooling between heating cycles (Figure 3.15).

Taken together, a pipette puller has a large number of parameters that can be set for each pulling cycle, and the number of cycles can be five or

3.3 ELECTRODES AND MICROPIPETTES

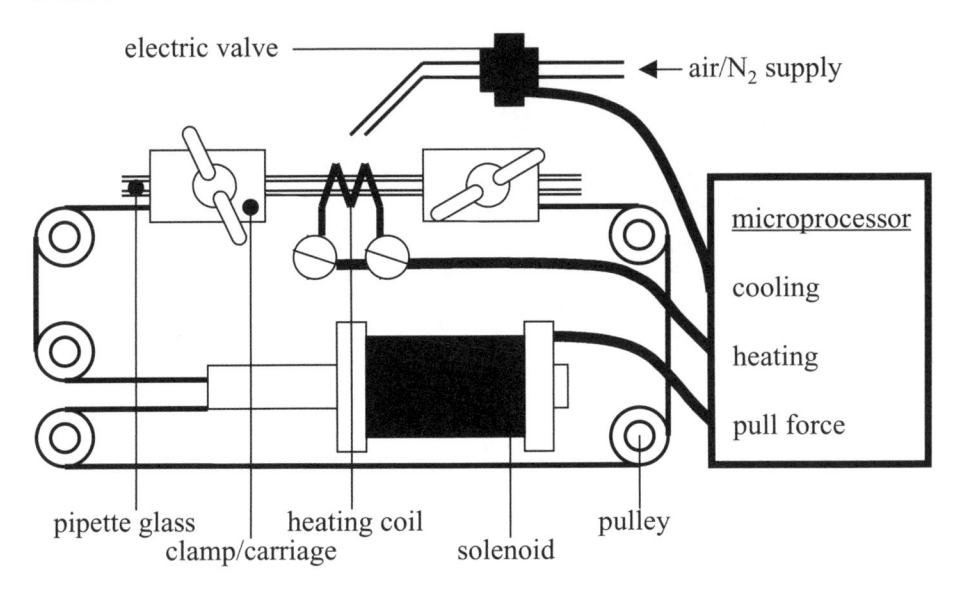

Figure 3.15 Elements of a pipette puller

more, although more than three are rarely used. Most pullers now have a small computer on board to manage these parameters (Figure 3.16).

Fire polishing of the pipette tip

A patch pipette is produced by a multicycle pulling process of which the last cycle is often a cold or semi-cold pull. As a result, the tip of the pipette is sharp and can have jagged edges. If such a pipette is used on a cell, it can easily destroy it. The problem is solved by fire polishing of the tip, i.e. briefly melting to make the edge rounded and smooth. Fire polishing can be done in some pullers as an extra cycle, but is mostly done separately. A home-made set-up or a commercially available device can be used. A set-up can be made from a microscope (with a 40× or 100× objective), a variable power source, such as an old transformer for a microscope light, and a filament (wire) mounted on a micromanipulator (Figure 3.17). The pipette tip is brought into the microscope field with the heating wire to which power is *gradually* applied. A minute change can be observed at the pipette tip, which is a sign that melting has taken place. The wire can be made of tungsten if that is available, but any very thin wire will do, keeping in mind that:

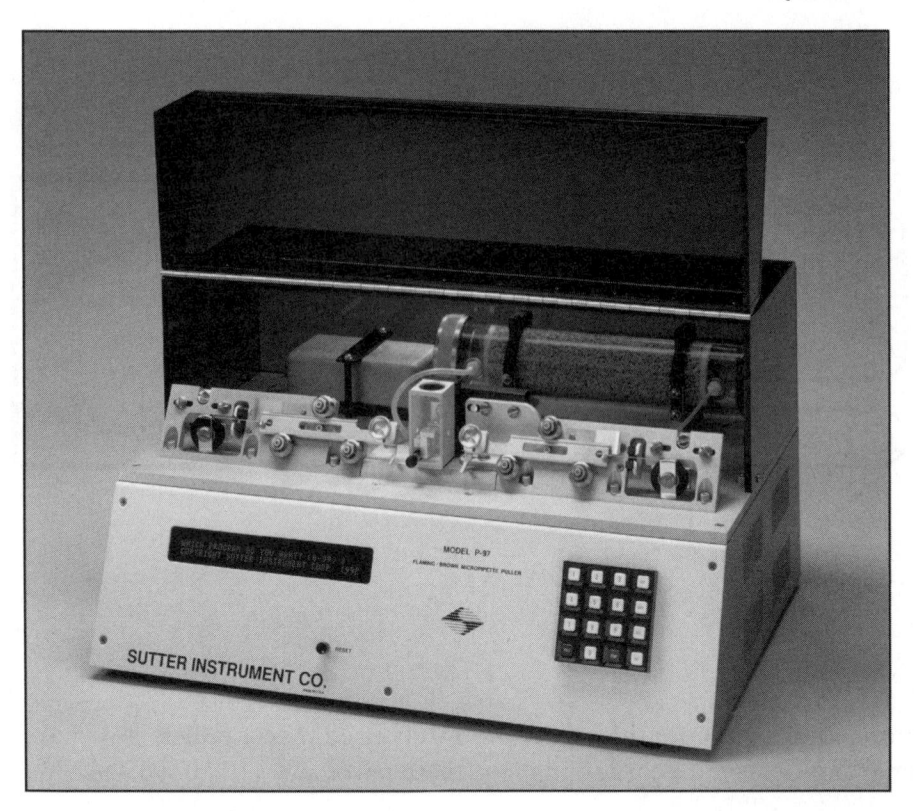

Figure 3.16 A commercially available pipette puller (reproduced with permission of Sutter Instrument Company)

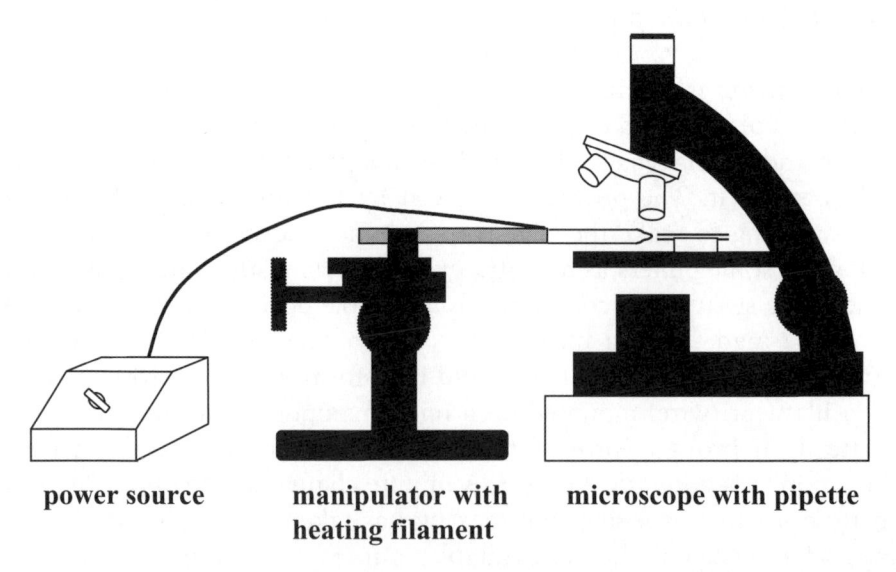

Figure 3.17 Principle of a fire polishing set-up made from common laboratory materials

3.3 ELECTRODES AND MICROPIPETTES

- The filament must be thin enough to act as a significant resistance.

- The filament must be much thinner than the leads from the power source (otherwise they will heat up as well).

- The filament must be thin enough to be visualised with the pipette tip in the same microscopic field.

- Extreme care must be taken in heating the filament in order to prevent melting, particularly if not using tungsten wire. This can be done by gradually increasing the current until the filament is red or orange hot. White hot is too hot! If a 100× microscope objective is used, it is wise to let the filament come up from an angle to minimise the chance of heat damaging the lens.

A commercially available fire polisher works in exactly the same way but does not require any DIY and provides very stable and precise control (Figure 3.18). A remark about patching in brain slices should be made here: in the experience of some workers, patching in slices works better

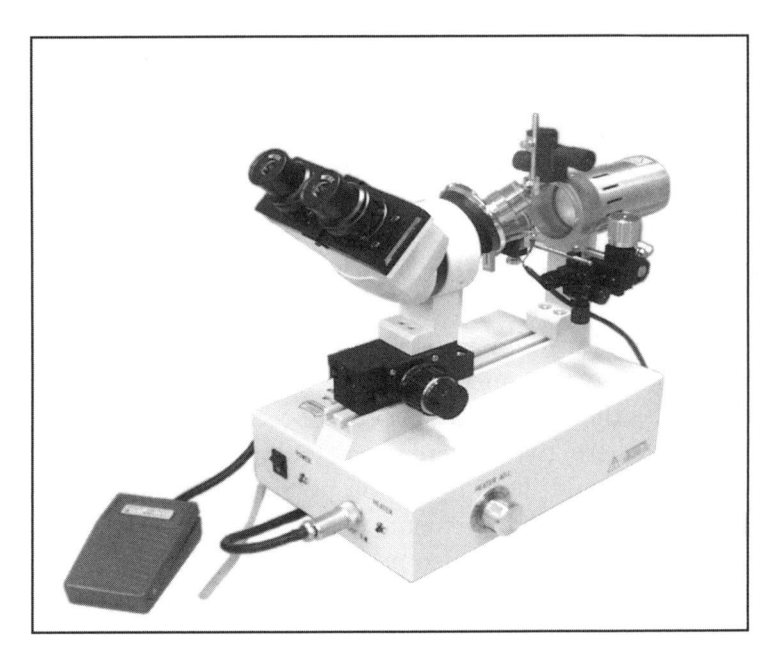

Figure 3.18 A commercially available fire polisher (reproduced with permission of Narishige Scientific Instrument Laboratories)

with pipettes that have had little or no fire polishing. This could be because fire polishing has a blunting effect on the tip, making it more likely to become contaminated while travelling through tissue than a sharper pipette. It is dependent on the pipette puller and the fire polishing practice, and requires experimentation.

Patch pipettes can be kept dust-free for days, e.g. in a petri dish held on a strip of plasticine. It is, however, better to fire-polish the pipettes just before use so that any contamination is burned off. Pipettes are filled from the back with a syringe and a long needle. Syringe filters should be used to ensure that no particles are introduced into the micropipette that can clog up the tip. The filament inside the glass lumen ensures that the fluid reaches the very tip. Filling needles must be very thin and long. Metal (HPLC) injection needles are sometimes used, but metal can dissolve in the solution, particularly if it contains EGTA, and can also damage the glass easily. A very cheap way to obtain filling needles is to pull them from 200 µl disposable pipette tips in a flame. Some skill and trial-and-error are definitely required here. Alternatively, nice non-metal, bendy filling needles are now commercially available, but pricey. The pipettes are filled from as near to the tip as possible. If small bubbles occur, then they can be shaken or tapped out. Unlike long intracellular recording electrodes, patch pipettes can withstand rigorous shocks. Pipettes are filled completely if the holder contains a silver pellet (the holder is then also filled). If the holder contains a silver wire, the pipette is filled to a level where the fluid makes good contact with the wire when the pipette is mounted in the holder. It is important here that the holder and pipette are not filled excessively because this can lead to unnecessary noise.

3.3.4 Liquid junction potentials

The final system of electrodes and solutions inevitably contains sources of off-set. In addition to the junctions between the solid (metal-containing) electrode and the salt solutions discussed in Section 3.3.1, another type of junction occurs where two different salt solutions meet. This type of junction is named liquid–liquid junction potential or simply liquid junction potential. The ease of movement of ions in solution is dependent on variables such as size, charge, temperature, hydration etc. and can be summarised by the term 'mobility'. Differences in mobilities and concentrations set up a potential between solutions, where ions tend to move along a 'mobility gradient'. It follows that if these differences are small, the liquid junction potentials are also small: in many standard situations less than 5 mV. The potentials are often ignored, even though it is not cancelled by symmetry (as with the solid–liquid junctions) or nulling the resting current before a cell is approached (see Section 4.1). As a consequence, problems arise in cases where high precision is required, or where vastly different solutions are used that can generate liquid junction potentials up to several tens of millivolts. It is always important to have at least some idea of the size and sign of the offset caused by the liquid junction potential and subsequent nulling by the experimenter. Most workers use (an adaptation of) the software developed by Barry (1994) to calculate the potentials. The programme is now incorporated in the pClamp software suite from Axon Instruments.

3.4 Electronics

In this Section the electronic requirements for patch clamping are discussed. Considerations roughly fall into the categories of noise reduction, data handling and experimental control. The latter two are illustrated diagrammatically in Figure 3.19.

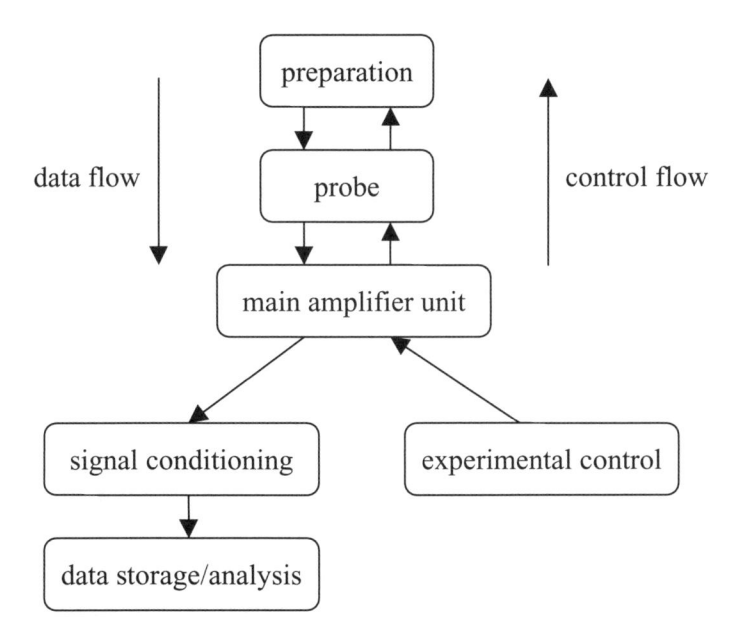

Figure 3.19 Diagram of control and data flow in a patch clamp set-up. The elements are discussed in detail in Section 3.4.2 and further

Patch clamp electronics can range from highly modular arrangements, comprising several devices controlled individually, to integrated systems where none of the devices have any controls and are slaved to a computer. In the following Sections the devices are discussed separately for clarity but in practice there is nearly always some integration. It will be signposted where necessary which options are fundamental and where common integration is found. The first category of considerations, noise reduction, has two aspects: reduction of disruptive external noise and handling of noise intrinsic to the system. The former is discussed below and the latter in Section 3.4.3.

3.4.1 External noise and Faraday cages

The order of magnitude of the currents recorded from single cells or single channels, together with the very high resistance circuits to record them, make patch clamping very prone to electrical interference. Any radio wave (in the broadest sense) will be picked up by the sensitive circuitry, particularly the probe, disrupting the experiment. The three most common types of external noise are:

hum **switch noise** **digital noise**

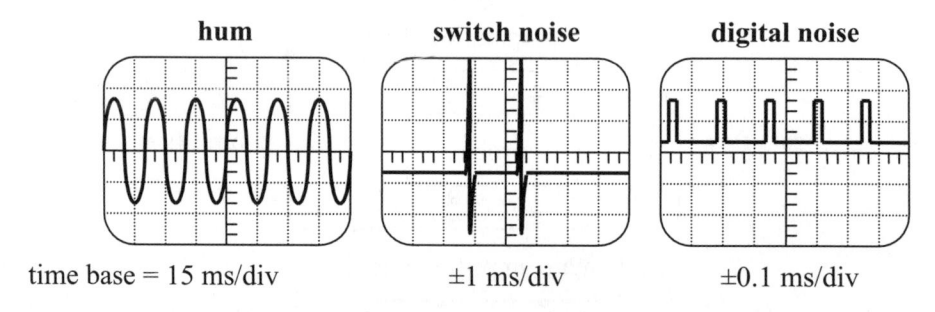

time base = 15 ms/div   ±1 ms/div   ±0.1 ms/div

Figure 3.20 The appearance on an oscilloscope of common types of external noise

1. *Hum*, so-called because it is a low humming sound when coming through a speaker. It is caused by the power grid alternating current, with a frequency of 50 Hz in Europe and 60 Hz in North America. The power grid is omnipresent and all active power leads can be viewed as hum-beaming antennas. Conversely, any conducting mass that is in contact with the 'hot' electrode and not grounded is a receiver

3.4 ELECTRONICS

of noise such as hum. Hum nearly always causes the biggest noise problems. It is solved by grounding objects and shielding sensitive circuits.

2. *Switch noise*. Devices that draw big currents and switch on and off create inductive drops and surges in the local power grid. These can appear as spike-like artefacts in recordings. The problem can be solved by surge protectors or by separating as much as possible the part of the grid used for items such as fridges and water baths from the part used for the patch clamp rig. Neither of these methods are necessarily 100 per cent effective and experimentation is often required.

3. *Digital noise*. Digital devices, particularly monitors, can be powerful sources of regular, high-frequency clock pulses. Keep them away from sensitive circuits.

This list is not exhaustive. For example, in medical schools the hospital radio suddenly can be heard through the patch amplifier speaker, or certain workers might have a talent for accumulating static electricity and create large, apparently mysterious surges in recordings independent of the rig they work on. Generally, however, external noise can be classified as radiation noise (entering the system as radio waves) or disturbances affecting the rig's power supply, and therefore can be treated like hum or switch noise, respectively.

Reducing radiation noise such as hum is done by removing sources (transformers and power cables are sources of hum) and shielding sensitive circuits with grounded metal objects. The primary circuit that needs protection is the recording circuit between the probe and the preparation. This is a very high resistance circuit with exposed parts such as the pipette and pipette holder measuring very small signals. The circuit has to have a high resistance to prevent loading (short-circuiting) of the biological preparation but because of the high resistance a small induced current in the circuit will result in a large voltage deflection (Ohm's law). Once the signal has been processed by the probe electronics, it is amplified and transported in low-resistance, well-shielded leads, and interference is much less likely.

(Electrical) ground

Shielding involves putting up radiation barriers in the form of metal objects that receive the noise and lead it to ground. Ground is therefore an important concept and deserves some discussion. Electrical 'ground' or 'earth' is a very large mass that effectively smothers noise by short-circuiting it via an ultralow resistance route. There is much variation in the quality of electrical ground, with the main variables being the ultralow resistance route and the size of interference it needs to handle. Ideally, ground is a large metal pole lodged deep into the earth, with only the rig's electronics and shielding connected to it. More often it is the earthing of the power grid. This can be very 'dirty' if the power cabling is long (creating resistance) and there are many appliances connected to it. It is sometimes better to create a private ground by using central heating piping or other metal building parts. If the quality of ground is not high enough, it will show through lower effectiveness of shielding and/or excessive surges coming through, and experimentation to find an alternative may be required.

Once the optimum ground is identified, it is imperative that devices are hooked up to it by very low resistance wiring, i.e. thick, short, copper wires. Long, puny ground wiring can annihilate a good ground by creating interference-prone resistances. If many items are connected to ground it is very easy accidentally to create *ground loops*. These loops will behave like induction coils and pick up interference. This must be prevented by drawing a diagram of the wiring and branching all ground connection from a single point (Figure 3.21). Most probes will have a separate ground input to be used for the bath electrode. This ground isolates and physically reduces the recording circuit considerably.

In some cases it is enough to connect all the metal objects around the preparation to electrical 'ground' or 'earth'. The microscope stage, lamp housing and micromanipulator combined can, depending on the environment, provide sufficient shielding for whole-cell recording. More often though, shielding in the form of a cage is necessary. A grounded metal cage (Faraday cage, Figure 3.22) excludes electromagnetic interference. Traditionally, Faraday cages are big affairs that cover the whole worktop and everything on it. They are usually home-made from aluminium or copper mesh mounted on a frame, with legs separate from the table to reduce mechanical interference. The inside can contain points where tubing systems are attached, and shelves to hold various devices. The cage can have doors that hinge sideways or upwards, although many doors on Faraday cages are never used because noise is already at acceptable levels without them, and closing them risks disrupting a patch. An alternative to the big cage is a small mesh that rests on the microscope platform and just protects the preparation and the probe. As mentioned, a big cage can also be used as a frame to support peripheral items, but it is sometimes difficult

3.4 ELECTRONICS

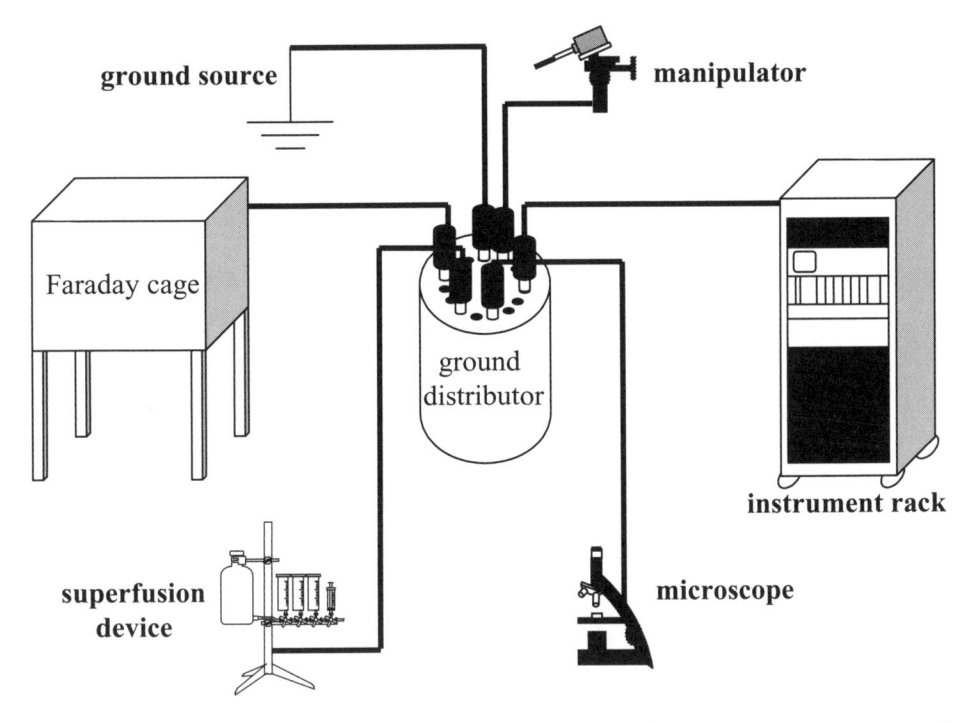

Figure 3.21 A ground distributor in the shape of a metal block or strip prevents ground loops

to keep all sources of noise, such as power supplies and electromotors, outside the cage, requiring long leads that in turn must be shielded. A small cage is much cheaper to make and easier to handle, modify and replace. From the above it is probably clear that I much prefer a small mesh cover on the microscope rather than a big, clumsy cage. In exceptional cases a big cage is necessary, e.g. if its roles include blacking out the microscope in fluorescence experiments.

Any necessary additional shielding of power cables, lamps, etc. can be done simply by wrapping the offending item in aluminium kitchen foil and grounding it using a crocodile clip (alligator clip in the USA). Care must be taken not to create ground loops here by resting the aluminium wrap on already grounded items.

Eliminating hum

Testing the rig for noise interference is done by applying a dummy load to the probe in the form of a model cell circuit or simply a resistor (470 MΩ

Figure 3.22 A commercially available, large Faraday cage (reproduced by permission of Technical Manufacturing Corporation)

or more) between probe input and ground. It is a good idea to deliberately make the connection between resistor and probe long (3–6 cm), in order to make the test circuit more sensitive to interference. Under voltage clamp, with a holding potential of zero, hum will be visible on the current output above the intrinsic noise (see Section 3.4.3). It is sensible to ground the main components of the rig before the apparatus is turned on, otherwise the hum level might damage the amplifier. The source of any remaining hum can be identified by touching the different components around the probe with a ground wire. If the hum does not change, then the item is not contributing significantly to the hum observed. (Note that possibly at more sensitive settings the item *does* contribute.) If the hum decreases, the item is definitely not grounded properly and needs a connection to the ground distributor. If the hum increases, a ground loop is created so the item is already grounded. The process can be repeated at increasing sensitivity until the hum level is 'acceptable'. What is acceptable depends on the application. A hum amplitude of 5 pA is not significant in whole-cell currents of several nanoamperes but will drown single-channel

3.4 ELECTRONICS 81

currents. It is nearly always possible to make the hum disappear in the intrinsic noise, the size of which is dependent on the filter settings (more detail on filters can be found in Section 3.4.3). Thus, it is important to use the amplifier and filter settings that are likely to be used in the actual experiments.

3.4.2 Patch clamp amplifiers

The apparatus that contains the measuring and clamping circuitry and controls is the central component of the set-up. It is usually a specialised piece of equipment that is purchased commercially, because it is not economically viable or at all possible to design and build such a high-specification device as a one-off in-house. Patch clamp amplifiers have a small extension or head-stage (probe) that is placed as close as possible to the preparation; the pipette holder is usually connected directly to it. The probe is the high-impedance (resistance) input of the amplifier and forms the interface between the main device and the preparation. Patch clamp amplifiers receive the data from the preparation and also process the experimental commands, such as holding potential and voltage steps. These commands can come from an external source or from devices integrated in the amplifier itself. Similarly, signal conditioning, which can include filtering, further amplification and data reduction, can be performed by external devices (see Section 3.4.3) or be integrated in the amplifier. Most commonly, some basic control commands (holding potential, simple pulses) and signal conditioning (gain, low-pass filter) can be performed by the amplifier, whereas more sophisticated tasks are done by separate devices. The most common features of the patch clamp amplifier are listed below:

1. *Current/voltage clamp selection.* A switch that toggles between the circuitry required for the two recording modes.

2. *Offset control.* A dial enables nulling of the resting current once a closed circuit is made, i.e. the pipette is in the bath.

3. *Capacitance compensation.* Capacitive transients caused by *RC* circuits (Section 4.2.1) as the holding potential changes can cause large artefacts. Most amplifiers have the possibility of generating one or more *RC* responses and subtracting them from the output signal, the cancelling the artefact. For each compensatory *RC* circuit both the

amplitude and the time constant can be set. There can be confusion over this feature caused by the misnomer 'capacitance compensation', which suggests that capacitive effects at the preparation are somehow reduced. This is not the case (in contrast to the effects of series resistance compensation, see item 4), and a better term for capacitance compensation is *capacitive transient cancellation* (Figure 3.23). In other words, the operation is purely cosmetic. On some amplifiers one RC circuit is designed specifically to cancel the access/pipette resistance – cell capacity RC – so that a good estimate of these parameters can be obtained and to feed the series resistance compensation circuitry (see item 4).

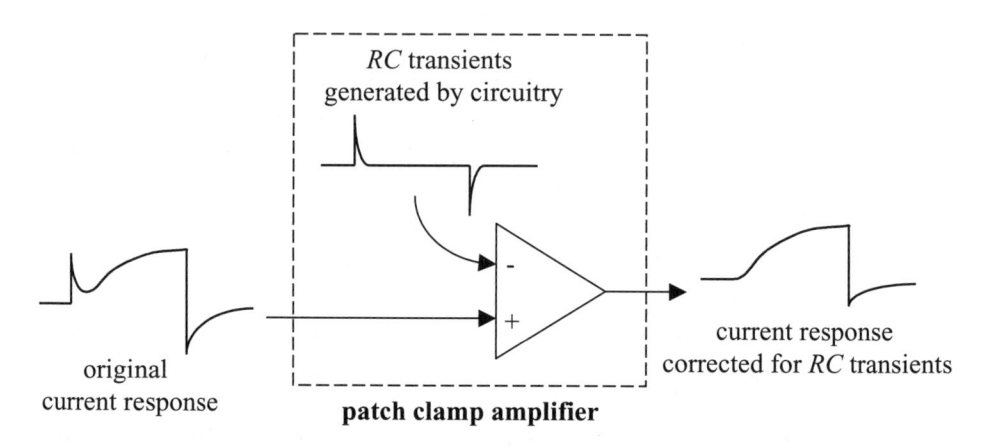

Figure 3.23 The principle of capacitive transient cancellation

4. *Series resistance compensation.* In the discussion of the equivalent circuit for the whole-cell mode (Section 2.3.6) it was mentioned that the main RC element in the recording circuit is formed by the access resistance R_{access} plus the pipette resistance $R_{pipette}$ (together named series resistance) in series with the cell capacity C_m. This 'series RC' can significantly affect the voltage clamp properties of the circuit, causing unwanted voltage drop and delays in voltage changes. The problem can be reduced in principle by over-injecting current when the holding potential changes, in order to speed up the response. The amount of over-injection can be related to the series RC time constant, which can be approximated on some amplifiers by cancelling the whole-cell capacitive transient (see item 3). A perfectly compensated series resistance would result in perfect voltage clamp of the cell

3.4 ELECTRONICS

membrane. However, the compensation circuitry is, by nature, very prone to hysteresis, i.e. self-amplifying destabilisation, usually in the form of oscillations. In practice about 80 per cent compensation is a very good result. It seems safe to avoid series resistance compensation altogether if the responses under study occur at a considerably slower time scale than the series RC time constant, but it must be kept in mind that a slow voltage clamp can lead to masking of effects such as activation, inactivation or desensitisation of factors that ultimately can influence the current under study, so controls with maximum attainable series resistance compensation can be worthwhile. In contrast to capacitive transient cancellation, series resistance compensation has a real effect on the voltage clamp properties of the system.

5. *Holding potential setting.* A multi-turn dial or digital setting enables the experimenter to set the holding potential. Depending on the specific amplifier, this can be combined with internal and/or external command voltage patterns and offset. Most amplifiers have a pulse oscillator to provide a regular voltage command pulse (e.g. 20 Hz with 20 ms pulse duration), to be used when establishing or checking the desired patch configuration (see Chapter 4).

6. *Gain.* A multiple switch to set the signal gain can be accompanied by a separate switch to set the gain of the probe circuit.

In addition to the above basic controls or their variations, amplifiers can have *low-pass filters* (see Section 3.4.3), *zap control* (a circuit to inject a very short, but large current to break the patch of membrane in order to establish the whole-cell configuration) and other useful or less useful controls.

Selecting an amplifier

In selecting a patch clamp amplifier there are many factors that must be balanced. Firstly, the application determines the amplifier specifications with respect to intrinsic noise, stability, probe size and gain. For example, there are now specialised amplifiers for low-noise single-channel recording and – on the other side of the spectrum – large-current oocyte recording (although the latter is not strictly patch clamping). Some amplifiers can be purchased with different probes to suit whole-cell or single-channel recording. Secondly, it is difficult to separate considerations for the amplifier from those concerning the rest of the electronics. It has been mentioned already that amplifiers can incorporate variable amounts and qualities of signal conditioning and command devices. The required specifications must be identified and taken into account in the decision. Generally, integrated systems are easiest to set up and use because all elements are designed to work together and the controls will be most ergonomic. However, a large investment must be made and the flexibility is as good as the manufacturer is able and willing to build into the system. A non-integrated system is flexible and can incorporate elements already available, thus saving money, but the ergonomics will not be optimal and there can be compatibility problems between devices, most commonly between the amplifier and the interface to the computer used for data acquisition and analysis. To summarise:

- Establish the requirements that are dictated by the application, not only for the amplifier but for all of the elements from Figure 3.19 (discussed in later Sections). Include likely future applications.

- Make an inventory of what equipment is available in your laboratory, if anything. Check compatibilities between devices.

- Get an overview of available makes and models and compare them on the criteria that are important to *you*, not those that are highlighted by the manufacturer.

- Put together a few systems that meet the requirements, attach the price tags and check your budget.

- Consider a balance of ergonomics, flexibility and cost (not forgetting service) for the whole system, and choose.

3.4.3 Noise prevention and signal conditioning

Signal conditioning is the process of modifying the recorded signals at appropriate points in the data flow to improve the signal-to-noise ratio, perform data reduction and provide compatibility between devices if necessary. To improve the signal-to-noise ratio, the external noise must be minimised as discussed in Section 3.4.1. More problematic can be the intrinsic noise, i.e. the noise inherent to the recording. A brief introduction to this noise is presented in the next section.

Intrinsic noise and its prevention

Noise in the context of electrophysiology can be defined as unwanted waveforms mixed with the recorded signal. Intrinsic noise is generated by the recording system itself. Most intrinsic noise is generated by thermal processes; the movement of atoms in a solid or a liquid due to thermal energy generates background electrical activity. In other cases, noise is generated by specific electronic components in the system or the source is not known at all. All of these types of noise are a superposition of waveforms of many different frequencies and strengths. The strength of a waveform or a group of waveforms can be quantified by the variance or square-root variance σ^2 (root mean square or rms). On some patch clamp amplifiers there is a facility to measure 'rms noise'. Noise is often classified by its *power spectrum*, which can help to pin-point the source. The power spectrum is built from the strength of the noise source (on the *y*-axis) at different frequencies (on the *x*-axis). The strength of thermal noise, depending on its source, can be proportional to the square of the frequency (f), just the frequency or be independent of frequency. The latter is referred to as *white* or *flat* noise. The most important non-thermal form of intrinsic noise is denoted by its power spectrum characteristic $1/f$, although the exponent of f is not necessarily exactly unity. This noise is not well understood and probably has multiple sources. Figure 3.24 shows the most common noise power spectra.

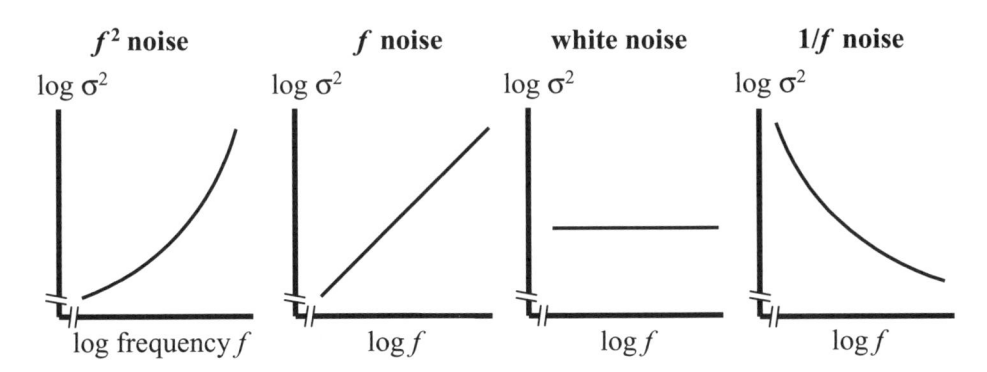

Figure 3.24 Power spectra of different types of common intrinsic noise

Thermal noise with an f^2 signature is associated with *RC* circuits, whereas thermal noise with an f signature is more specific to the dielectric properties of capacitances only. White noise, if thermal in origin, is *inversely* proportional to resistance. Although the physical mechanisms of noise generation are not always understood, the experimental factors involved are known and when troubleshooting or when ultralow noise recording is required, these can be tweaked. For example, in sensitive single-channel recording the capacitances formed by the micropipette and possibly the pipette holder can become very significant. These capacitances contribute to thermal noise through RC noise and dielectric noise, and can be minimised by making the pipette and holder short (reducing the capacitor plate area) and/or thick (increasing the capacitor plate distance). A special case of capacitance prevention is to thicken the micropipette wall after pulling by application of an insulating layer of beeswax or silicone gel. This reduces pipette capacitance by increasing capacitor plate distance but also by preventing a water film from creeping up the outer pipette wall to create more capacitance. The thickening process – or 'sylgarding', named after the silicone product that is often used for it – can be tricky because most benefit is gained by treating the pipette close to the tip where the glass is thinnest. It is sometimes more practical to direct effort towards pulling micropipettes with very blunt (fast-tapering) tips.

From the above it is evident that intrinsic noise is not just a nuisance but also contains information that might be useful, such as indications of the number and activity of ion channels contributing to a current. This property is still important for studying very small conductance channnnels. Although noise analysis is not discussed any further here, it still makes interesting reading and has applications (e.g. Chung and Pulford, 1993).

Filters

Minimising the generation of intrinsic noise is not enough. In most instances filters are necessary to process the recording in order to extract the relevant signal. Before electronic filters are discussed, it is appropriate here to establish a very important principle: filtering leads, by definition, to loss of information. Taken to the extreme, data can be stored best in its rawest form, so that no information is lost. Filtering then can take place at will in the analysis process *off-line* (after the experiments). However, the nature of storing devices, both analogue and digital (see Section 3.4.4), limits the amplitude resolution of stored data, so that recordings with a very poor signal-to-noise ratio yield poor results when analysed off-line. In other words, high noise levels can affect the resolution of the recorded signal. This consideration is relevant if the noise level is considerably higher than the signal amplitude. In practice, recordings need some

3.4 ELECTRONICS

filtering before they are stored, but choosing the filtering should be based on maintaining signal resolution instead of recording the prettiest pictures you can visualise at the time.

The term 'filter' in the context of signal conditioning refers to the filtering of frequencies. A recording – being the sum of noise and the signal(s) of interest – can be characterised, like noise, by a power spectrum. If the range of frequencies that includes the signal of interest is known, then all other frequencies can be filtered out. There are four main types of filters classified by the filter characteristic. The filter characteristic is a power spectrum that indicates the output of the filter when the input is white noise (flat power spectrum). For example, a low-pass filter (Figure 3.25) will show high power at low frequencies (indicating little or no resistance) and increasingly stronger blockade at higher frequencies. Low-pass filters are most commonly used because biological signals usually range from DC (frequency of zero) to an upper frequency limit, whereas noise has a much wider frequency range. High-pass filters are used in studies of fast events and they eliminate – or rather mask – slow drift. Band-pass filters are a combination of a low-pass and a high-pass filter to leave a well-defined range of frequencies, whereas a notch filter selectively blocks a range of frequencies. The latter is often used to filter out hum.

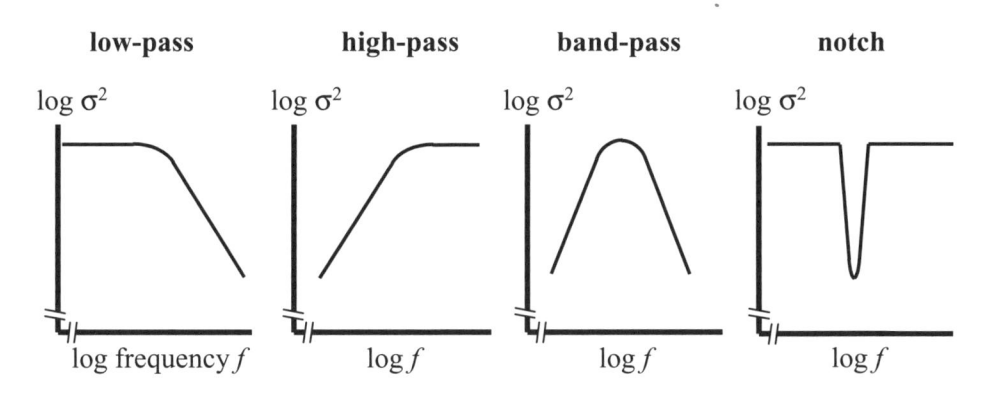

Figure 3.25 Filter characteristics are power spectra from the outputs of filters when the input is white noise

In addition to the shape of the filter characteristic, a filter is defined by its *cut-off frequency* and *slope*. For example, a low-pass filter will cut off all frequencies higher than a value set by the experimenter, and the steepness of the slope down towards higher frequencies in the filter characteristic is a measure of how effective the filter is. In general, but not always, the cut-off frequency of a low-pass or high-pass filter is defined as the frequency where the *power* ($\equiv$variance or σ^2; note that power does not equal amplitude) of the original signal is dampened by -3 decibels (dB) or about half the original strength. For a sine wave, this corresponds to a reduction in *amplitude* to $1/\sqrt{2}$ or 71 per cent of the original, which implies that dampening occurs on either side of the cut-off frequency, unlike the widespread misunderstanding that the cut-off frequency marks the starting point of dampening.

The steepness of the characteristic can be expressed in dB per decade, i.e. the dampening ratio between a power of ten units of frequency. *Orders* or *poles* are used as units to indicate steepness: a first-order low-pass filter is -20 dB per decade, a second order filter is -40 dB per decade, etc. Alternatively, steepness is expressed in dB per octave; the dampening ratio between a power of two units of frequency (see Figure 3.26 for an example). It is imperative to be aware of the filter properties in your system in order to avoid *aliasing* (see analogue-to-digital conversion, Section 3.4.4) or losing important data.

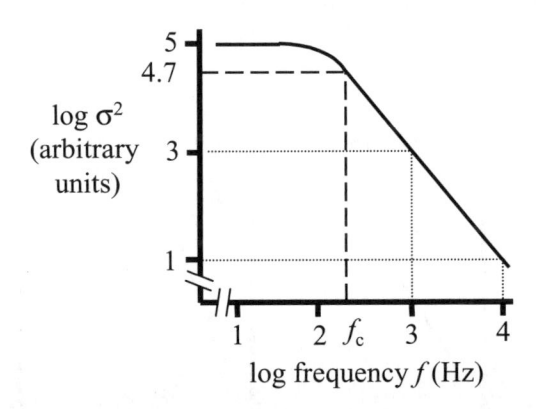

Figure 3.26 Filter characteristic of a first-order (-20 dB per decade) low-pass filter with a cut-off frequency of 200 Hz

Different (analogue) filters have special characteristics that can be important in data analysis and interpretation. Bessel, Butterworth and Tschebycheff filters have different responses to waveforms: the Bessel filters do not create additional oscillations but the other two have steeper characteristics. The ideal filter is a Gaussian filter. It can be modelled exactly but can only be approximated physically by electronics. Another consideration regarding filters involves phase shift. Physical or analogue

3.4 ELECTRONICS

filters can only react to a signal coming in, so they necessarily create a time delay in the waveform (Figure 3.27). Both the problems of filter characteristic definition and phase shift can be solved by off-line digital filtering. This involves applying an algorithm to a digitised recording. The algorithm for a perfect Gaussian filter is simple and is part of most analysis software. Phase shift can be prevented by adjusting the time-base. These superior off-line filter qualities are another reason for recording data as raw as possible.

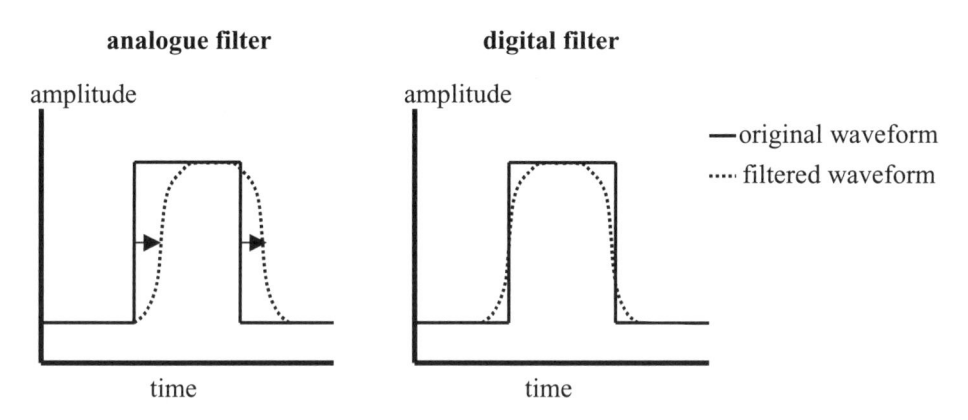

Figure 3.27 Analogue filters, by their nature, induce a time shift (phase shift) that can be prevented with off-line digital filtering

Other forms of signal conditioning

The most obvious signal conditioning is amplification. In electrophysiology, recordings contain very small signals that need to be amplified before they can be processed further. In addition, owing to biological requirements, the recording circuit, which includes the preparation, must be of very high impedance (resistance). Such a circuit is very prone to interference, so one of the first signal processing steps is impedance conversion. Both initial amplification and impedance conversion take place in the probe, as close as possible to the preparation (see also Section 3.4.2). Further amplification is performed in the patch clamp amplifier, mainly to scale the recording amplitude to the input range of the analogue-to-digital converter (see Section 3.4.4).

If the data of interest can be identified using the passing of a threshold, such as by the opening of a single channel or the occurrence of an action potential, data acquisition can be restricted to those events. Thresholds can be set by the experimenter after the data are digitised or before that, using an *event detector* or *discriminator*. If event detection is used in the recording stage then large data reductions can be accomplished, but the experimenter must be confident that the detector is flawless!

3.4.4 Data acquisition and digitisation

Early patch clamp data were often recorded on analogue media such as video recorders. In some laboratories remnants of this can still be seen in the form of analogue recorders or their modern equivalents, DAT recorders, now used to provide a slow 'overall recording' that allows the experimenter to monitor and possibly later re-examine the experiments over a long time scale. Gradually data acquisition has evolved from using household appliances via digitisers and mini computers to on-line personal computers. Personal computers are also used to perform experimental commands and data analysis. To provide commands and acquire data, signals must be converted from analogue to digital (AD) format, and vice versa. The device to perform these tasks is the AD/DA converter or interface. Once analogue data have been digitised and stored, they are no longer subject to deterioration. Unlike analogue signals, digitised data can be communicated without distortion and processed by – in principle – infinitely flexible computer programs. The AD step, however, is critical and deserves special attention. Digitisation involves dividing continuous data into discrete numbers by sampling. The quality of conversion is dependent on optimalisation of temporal and amplitude resolution:

1. *Temporal resolution in AD conversion.* An analogue signal consists of an infinite number of data time-points. Analogue-to-digital converters sample this signal at regular time intervals. The interval time is set by the experimenter and is a trade-off between data reduction and the quality of the digital recording. If the interval time is too long, data will be distorted or artefacts can be introduced (Figure 3.28). The latter phenomenon is called *aliasing*. The sampling interval time is set through computer software and is sometimes expressed as the sampling rate (samples per second, i.e. the reciprocal of interval time). The choice of sampling interval is not always easy and is strongly dependent on the signal of interest and the required analysis. Most whole-cell recordings of voltage-dependent currents can be covered adequately by 1024 or 2048 samples per trace, but receptor-operated channel activity, which can be slow to develop but requires good

3.4 ELECTRONICS

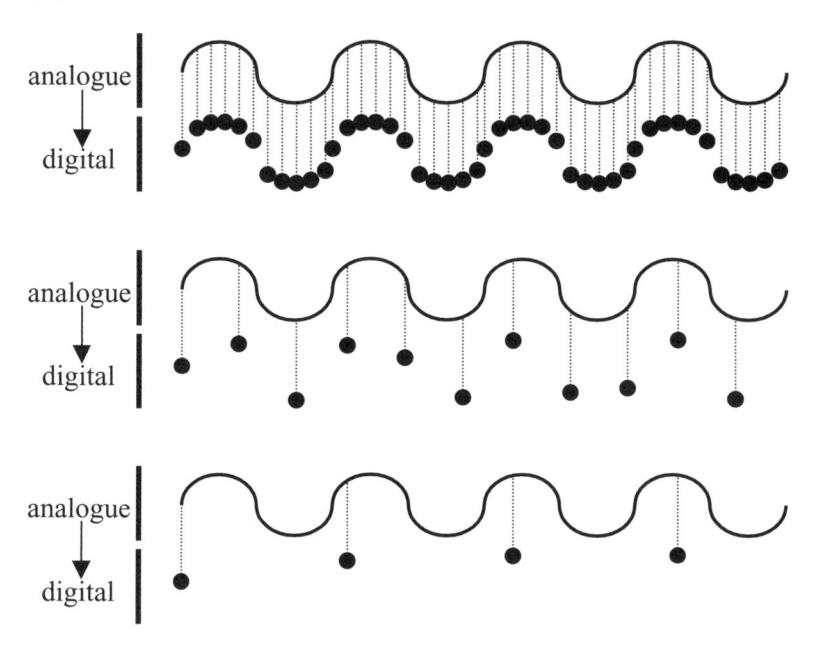

Figure 3.28 Effect of different sampling rates on quality of rendition of an analogue signal. The top panel shows sampling that results in a reasonable representation of the analogue signal. The low sampling rate in the middle panel results in a severely distorted digital signal. The bottom panel demonstrates aliasing: the appearance of artefacts in the form of new (slower) waveforms when the sampling rate is grossly inadequate

temporal resolution, is less straightforward. Some workers resort to using DAT recorders for these situations (after all, a sampling rate of 44 kHz is very decent for most applications).

2. *Amplitude resolution in AD conversion*. Amplitudes sampled at discrete time points are stored in digital format. The number of bytes used to store a number determines the range of numbers available. For example, two bits can combine in four different combinations: 00, 01, 10 and 11. In general, there are 2^n possible combinations for a group of n bits. Consequently, if the input range of an AD converter is specified to be '± 10 V, 10-bit', then the analogue input range of -10 to $+10$ V is subdivided into $2^{10} = 1024$ sections. Most AD converters have 16- or 24-bit inputs that allow about 65 thousand and 17 million sections, respectively. Although this seems generous, it is important that the analogue signal gain is set so that this range is used optimally. If a very small signal is offered at the input of an AD converter, only a

fraction of the possible amplitude resolution is utilised. For this reason the gain of the patch clamp amplifier should always be set so that the signal covers most of the AD input range. If the signal is outside the AD input range, then *clipping* occurs: the converter will store that data as maximum or minimum amplitudes, thereby cutting the signal (Figure 3.29). If an oscilloscope is used to monitor the signal, it is a good idea to set the amplitude range to the AD input range and not touch this setting during the experiment. This increases the chance that the signal is scaled properly for digital storage.

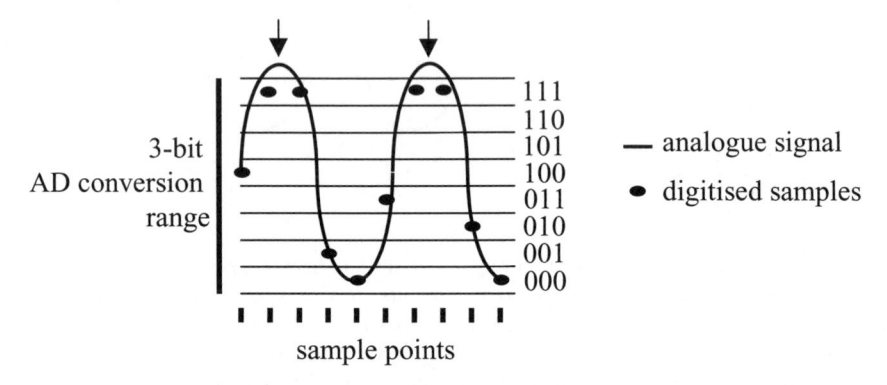

Figure 3.29 Example of a 3-bit AD conversion. If the analogue signal is outside the AD input range clipping occurs, as indicated by the arrows

The commonest data stored in patch clamp experiments are membrane currents, but sometimes there is another AD conversion channel employed for the command voltage/holding potential combination as well. This is not as superfluous as it may seem, because the holding potential can be set on the amplifier as well as on the computer and the value on the amplifier will not be stored with the recorded data. In addition, on some amplifiers the voltage signal reflects the actual voltage as 'seen' by the preparation, which might be different from the programmed voltage, particularly at times when the voltage is changing.

Experimental commands such as command voltage are communicated from the computer to the patch clamp amplifier through the process of digital-to-analogue (DA) conversion. The interface that performs AD conversion usually also has several DA channels. Additional features, depending on the system used, might include AD conversion channels that automatically read amplifier settings such as gain, and digital (TTL) channels that can be used for timing devices and triggering oscilloscopes.

3.4.5 Computers and software

The functions of the computer and software are:

- to control and monitor the experiment;
- to store data;
- to perform data analysis.

Current personal computers are easily powerful enough to communicate with AD/DA interfaces for most patch clamp applications and to perform complicated data analysis. The personal computer therefore provides a very powerful, accessible and uniform tool in the patch clamp set-up. The software can be a loose set of modules performing different tasks or an integrated package. The quality of the packages, at least from the main manufacturers, has consistently improved at a high rate over many years and the choice of package is usually determined more by the choice of AD/DA interface than by features of the software. In most cases the software is associated with at least the AD/DA interface because the two need to be directly compatible. That is not to say that there are no other differences between products. Although most packages will perform complex pulse protocols, data acquisition and analysis, there are differences in advanced features and user-friendliness that can tip the balance. For example, market leader Axon Instruments presents a very good Windows^{TM}-based package that integrates the software with the AD/DA interface and the patch clamp amplifier. Unfortunately, taken together – which is practically inevitable – this ensemble requires substantial investment. (As with all purchasing considerations, budget restrictions can defeat any other argument.) Other AD/DA interface manufacturers, such as Cambridge Electronic Design, provide a more fluid software product where it is possible for users to write routines themselves to suit any application. Other manufacturers might boast yet other special features. Similar to patch clamp amplifiers, the final choice of software and AD/DA converter is dependent on application, interconnectivity with other components of the rig and price.

4
The Practice of Patch Clamping

In this Chapter the actual procedure of making a gigaseal on a cell membrane and establishing the desired configuration is discussed. The purpose of this is to indicate good practice. Practitioners might find that this Chapter helps to increase the success rate, facilitates trouble-shooting and reduces having to 'reinvent the wheel'. Novices hopefully will be saved much time. The method used in this Chapter is to provide a step-by-step account of how to establish a patch clamp configuration, using the background information outlined in the previous Chapters. Electrophysiology is one of the disciplines within biology that allow many instant quality controls; these will be indicated where appropriate.

4.1 Preparing the Experiment and Making a Seal

The process up to establishing a seal between the membrane of the cell under investigation and the patch pipette is common to all forms of patch clamping and will be described in the following Section.

4.1.1 Setting up

The start of a patch clamp experiment, assuming that the set-up is complete and operational, usually consists of making solutions and turning on any water baths and thermo-elements, the heating of which can take some time.

Salt solutions

The solutions commonly mimic the ionic composition of the physiological fluid that the plasma membrane is exposed to *in vivo* or in culture. Thus, 'extracellular solution' (ECS) is used as the bathing solution and consists mostly of sodium chloride. 'Intracellular solution' (ICS) used to fill pipettes for whole-cell and outside-out excised patch recording contains much potassium. Typical ECS and ICS compositions are presented in Table 4.1. Perhaps even more than ionic composition, of which often more than one are used in a series of experiments, pH and osmolarity are critical to success in obtaining good seals.

Table 4.1 Examples of typical extracellular and intracellular solutions (ECS and ICS) used in patch clamping a mammalian cell. Note that the pH is set in ECS using NaOH and in ICS using KOH, the volumes of which can be significant in determining the reversal potential for sodium and potassium, respectively

Chemical	ECS concentration (mM)	ICS concentration (mM)
Na^+	126	5
K^+	6	147
Mg^{2+}	2.5	1.2
Ca^{2+}	1.2	0
Cl^-	125	150
GTP	0	0.1
ATP	0	5
HEPES	10	20
Glucose	11	11
Sucrose	67	0

Logically, ECS should have the pH and osmolarity of the environment the preparation is taken from, e.g. the culture medium in the case of cultured cells. The pH is controlled by a buffer, of which several types can be found in the patch clamp literature. A carbonate buffer consisting of an equilibrium between gaseous CO_2 and HCO_3^- is considered by many to be the most 'physiological' for vertebrate preparations but its use necessitates constructions to keep the solution exposed to CO_2 – or rather not exposed to air – at all times. Other buffers such as phosphates or HEPES, although widely used, have been reported to have side-effects in some instances. Common sense ultimately decides what to use. It is difficult to object to using a HEPES-buffered ECS for cells that have been growing in HEPES-buffered medium, but it is safer to use carbonate-buffered ECS for freshly dissociated cells or tissue slices. Buffering pipette solutions is often

4.1 PREPARING THE EXPERIMENT AND MAKING A SEAL

more problematic because the gas the solution is exposed to is normal air, which excludes carbonate buffering. Furthermore, ICS often requires quenching of calcium ions by chemicals such as EGTA, which makes the solution very acidic, and higher concentrations of HEPES or other buffers are required, amplifying any unwanted effects. Most vertebrate ECS is set to the physiological pH of about 7.4 and it is important to be consistent throughout a series of experiments because even small changes in acidity can influence channel behaviour. ICS is a little more acidic at about pH 7.2–7.3.

The importance of osmolarity is often underrated. Both absolute osmolarity and the difference between solutions on either side of the membrane are very critical for success. Osmolarity differences between the media on either side of a cell membrane determine the volume of the cell and the osmotic force on the membrane. If the difference is large, the cell will die. Small differences, however, can be exploited to increase success. For example, if the intracellular osmolarity is a little higher than the extracellular osmolarity (say 340 versus 330 mOsmol dm^{-3}), cells with wrinkly membranes will swell slightly so they can be patched better. Similar to pH, it is logical to set the ECS osmolarity to that of the medium that the preparation originates from. This can lead to very high extracellular ion concentrations and possibly undesirable equilibrium potentials (see Section 2.2.1). Some workers use relatively inert sucrose to obtain the necessary osmolarity while keeping the ion concentrations low and constant. Osmolarity theoretically can be calculated as parts per volume of solvent, but much chemical data regarding weak bases and acids, etc. must be known and processed. Osmometers provide quick readings, allowing comparisons and quality control.

It is practical to keep a stock bathing solution of tenfold strength and make up the desired quantity on the day. Calcium and glucose are often omitted in stocks to prevent precipitation of calcium salts and growth of microorganisms. Intracellular solutions often contain perishable chemicals such as ATP and are made in a batch for aliquotting. The aliquots are kept frozen.

There are at least two processes occurring during the preparation stage that by their nature, progressively worsen the experimental conditions over time. These processes are deterioration of the preparation and contamination of the pipette tip. There might also be others, such as the breakdown of drugs. It follows that the preparation should be placed in the set-up at the last possible moment, taking into account any equilibration time that the preparation might need. An important decision at this stage is whether or not to switch on the bath perfusion. On the one hand, an argument against constant superfusion might arise if the bathing solution contains expensive drugs. On the other hand, carbonate-buffered solution will become alkaline when exposed to air, so if carbonate-buffered solution is used then constant superfusion might be obligatory. In addition, if during the planned experiment drugs need to be administered by superfusion, it is a good idea to have the superfusion flowing before making the seal, in this way the flow can be switched to drug-containing solution instead of having the mechanical disruption of starting and stopping the flow during the experiment.

The patch pipette should be made and used immediately to reduce tip contamination and subsequent bad sealing properties. The critical step here, as mentioned in Section 3.3.3, is the fire-polishing step that cleans the tip. Pipettes can be pulled and stored dust-free in batches so that they only need fire-polishing and filling just before use. After the pipette is prepared and mounted in its holder, the tip is positioned above the bathing solution, straight over the preparation.

4.1.2 Bringing the pipette near the preparation

Before lowering the pipette into the bathing solution, there must be a slight pressure on the pipette fluid to blow any contaminations in the bathing solutions away from the pipette tip. These contaminations often gather in the bathing solution at the fluid-air interface, so pressure must be on before this is crossed. A pressure of about 10 cm of water is usually enough. Extreme pressure can affect or even blow away your cells! This is also the time to apply a small oscillating test pulse, in voltage clamp or 'search' mode. The current response to a small block pulse (typically 5 or 10 mV, 30 ms, 20 Hz) is an important aid to guide and check the process of making a seal (Figure 4.1). A simple pulse oscillator is often found on the patch clamp amplifier or is part of the computer software. When the pipette is outside the bathing solution, no current will flow because the resistance in the circuit is, for the purposes here, infinite. Some fast peaks might be visible at the start and the end of the pulse as a result of stray capacitance within the probe.

When the pipette is lowered into the bathing solution using the coarse manipulator, the current changes to a large square response, signifying the pipette resistance. The resistance can be calculated easily by dividing the test pulse amplitude by the current response amplitude, taking care to measure points well away from any capacitive transients (see Figure 4.2).

4.1 PREPARING THE EXPERIMENT AND MAKING A SEAL

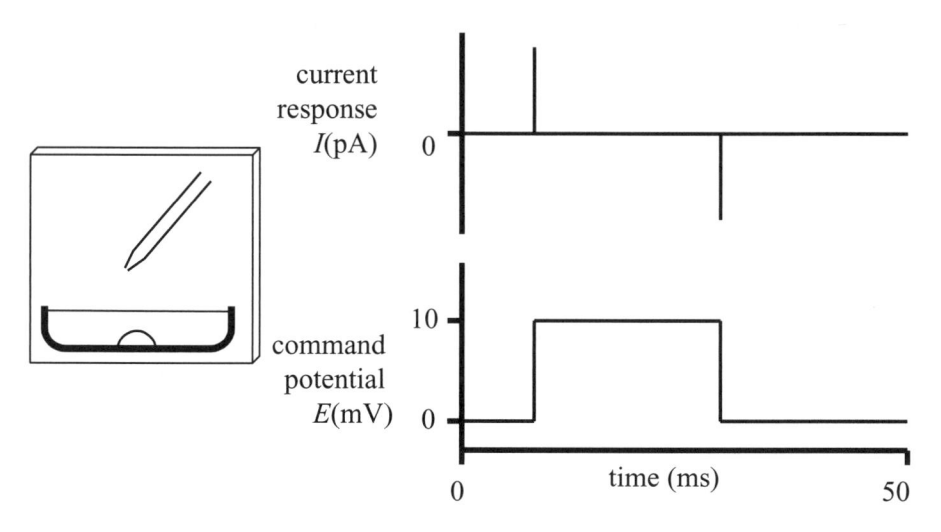

Figure 4.1 Current response to a block pulse with the patch pipette out of the bathing solution

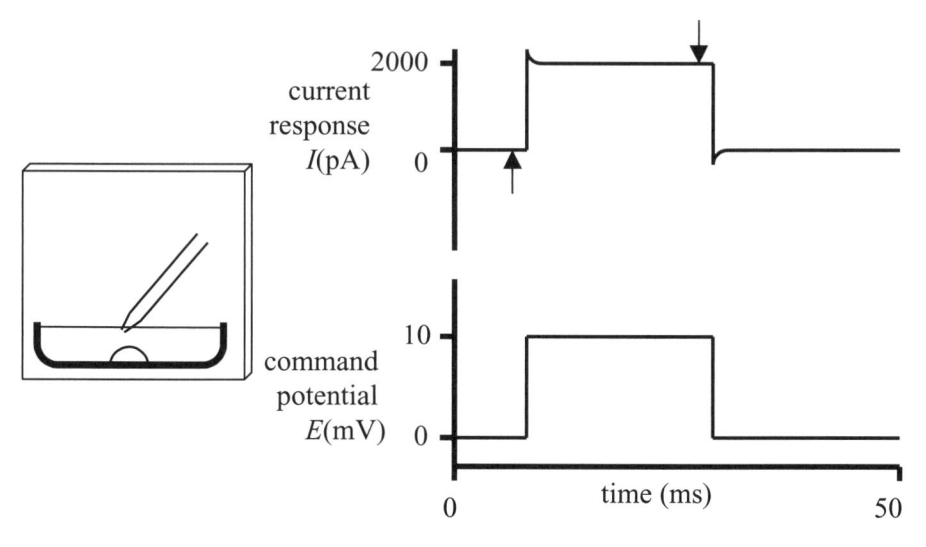

Figure 4.2 Current response to a block pulse with the patch pipette in the bathing solution. The response indicates that the pipette resistance $R_{pipette}$ is 5 MΩ. Arrows indicate where measurements should be taken to avoid the capacitive transients

In some software packages this operation is performed automatically and continuously.

It is often easier to monitor this response than to try to see the pipette tip entering the fluid. At this stage it is immediately clear whether there is something seriously wrong with the pipette. Some common faulty responses are:

- A very noisy signal but no square response: there is no ground electrode.

- A very large block response: the pipette tip has broken off.

- No change even though you are sure the pipette entered the bathing solution: the pipette fluid does not make contact with the electrode or there is an air bubble in the pipette tip.

- A normal pipette response but followed by a sudden drop in amplitude, i.e., increase in resistance: contamination material in the pipette fluid is pressed into the tip.

If all is well, the tip must then be guided towards the cell under study. This requires practice and you are very likely to break some pipettes when getting to know a new rig. You have to be acutely aware of the three-dimensional movement of the manipulator, both coarse and fine, in response to handling the controls. You should also know at any moment where the focal point of the microscope is situated. In approaching the cell, the focal point should be between the cell and the pipette. (If you try focussing on the pipette tip at this stage you might force the objective through the bottom of your bath.) Initially the tip is identified by moving it in the x–y plane. It is recognisable by a vague, dark shadow, the tip of which should be in the centre of view. The pipette then can be lowered carefully until the tip comes into focus. During the lowering, the tip might wander in the x–y plane if the axes of the coarse manipulator are not exactly perpendicular to the microscope light path. From then on it is a matter of lowering the focal point and the pipette position iteratively until the cell comes into focus. Experienced workers can do this in one step. During this process the current response to the oscillating test pulse should not change. The pipette tip is placed in a position where movement along one axis of the fine manipulator (usually the longitudinal axis of the pipette) will bring the pipette into contact with the cell at the desired location.

4.1.3 Making the seal

In the final position, contact between the pipette tip and the cell should not stress the membrane but be complete all around the tip. It follows that the spot to aim for on a cell should face the pipette tip as straight on as possible, and that the final movement of the pipette should be as near the longitudinal axis of the pipette as possible (Figure 4.3).

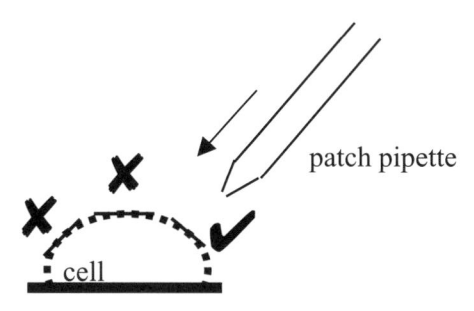

Figure 4.3 To obtain full contact between the cell and the patch pipette without stressing the membrane, the approach and location must be considered carefully

Once the target is chosen, the pipette tip is positioned near the cell using the fine micromanipulator. At this point the final baseline nulling takes place using the amplifier's offset knob, because this is the last stage where the recording apparatus is not influenced by the cell. Ideally, the final approach is monitored simultaneously on the microscope and on the oscillating test pulse. This is possible if the microscope is equipped with a video camera and the video monitor is next to the oscilloscope or computer monitor. The transition between final approach (as seen on the video screen) and contact (as indicated by changes in the test pulse response) can then be executed fluently. However, most patch clamp rigs do not have a video camera, so the experimenter must use judgement and look back and forth between the microscope and the test pulse to perform this vital step. (In the case of recording from a cell in tissue such as brain slices, the experimenter often cannot see the exact cell that is going to be patched at all.) Near-contact is indicated by a dip in the test pulse response (signifying increased resistance) caused by occlusion of the pipette opening by the cell (Figure 4.4). In some cases the response also wobbles because the pipette pressure causes the membrane to flutter slightly.

The optimum size of increase in resistance is variable from situation to situation, depending on pipette geometry, cell type and pipette pressure. In

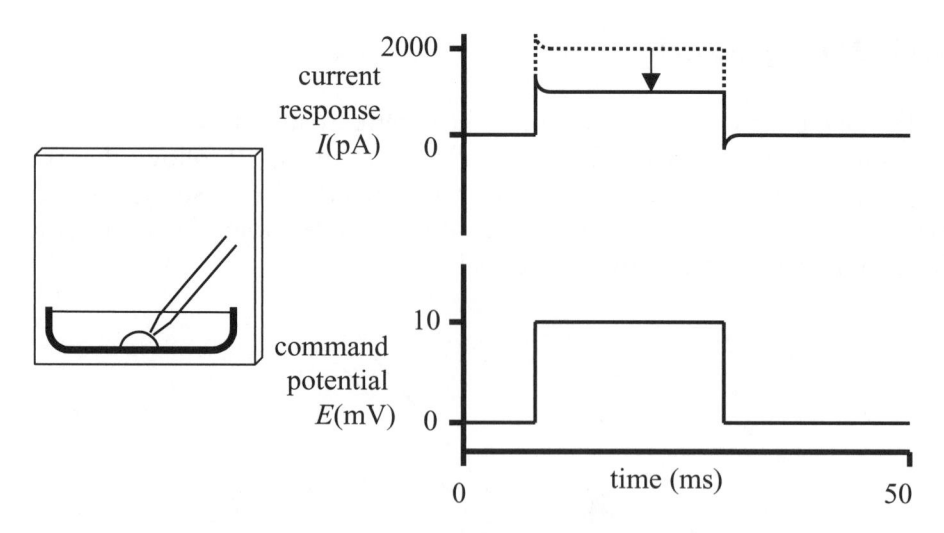

Figure 4.4 Current response to a block pulse with the patch pipette nearly touching the cell. The response drops in amplitude, indicating increased resistance

most cases an increase of around 1 MΩ is right. From this stage the pipette should not be moved anymore, because the tip can be, at least in part, in contact with the cell membrane and must be considered 'contaminated'. The next step is to remove the pipette pressure, which should drop the test pulse response considerably. If not, it is very unlikely, but not impossible, that a gigaseal can be obtained. The gigaseal can establish itself spontaneously but unfortunately in some preparations this is a bad sign: the pipette is pressing on the membrane, making it liable to rupture on the next manipulation, e.g. attempting to establish the whole-cell configuration (see Section 4.2.1). Ideally a little suction – applied by mouth – will give the desired result. There are a few issues regarding this stage in the experiment that are worth mentioning:

- If during this phase the test pulse response increases significantly (reduction in resistance), then there is practically no chance of obtaining a gigaseal after that. The pipette tip will have cell material on it, which shields some of the glass surface from the contact area on the cell. This makes it impossible to form a complete seal. The only thing to do is to replace the pipette and start over again. This may seem laborious, but it is quicker than the countless hours wasted by inexperienced experimenters trying to obtain a seal with a contaminated pipette.

4.1 PREPARING THE EXPERIMENT AND MAKING A SEAL

- Related to the above is the practice of removing the pipette pressure and observing the test pulse response for a while before applying suction. This creates the possibility that the membrane will 'wobble' and contaminate the pipette tip. It is best to remove the positive pressure and immediately apply a little suction in a smooth transition.

- Conversely, as long as the test pulse response is constant or decreasing, there is the possibility that sealing will be successful. Some cells require much patience and/or considerable suction. Generally, the attempt is over when the test pulse response is increasing (seal break), when a capacitive transient appears (the membrane patch under the pipette tip has ruptured) or when a gigaseal is established (Figure 4.5).

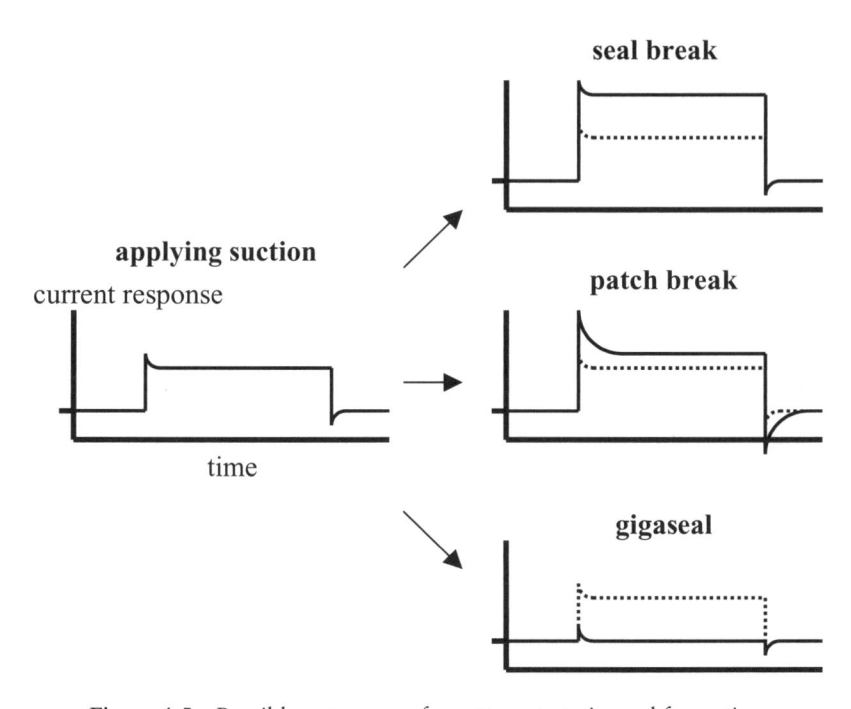

Figure 4.5 Possible outcomes of an attempt at gigaseal formation

A successful seal is indicated by a virtual absence of current response to a test pulse, as indicated in Figure 4.5. The resistance should be more than 1 GΩ for most whole-cell recording applications, and even better for single-channel recording. If the seal procedure has been performed in 'search' mode (where offset is automatically corrected), then now is the time to switch to voltage clamp. If the patch seems stable (low noise level) and the pipette tip does not drift, then all is ready for the next stage: establishment of the configuration of choice.

4.2 Whole-Cell Modes

4.2.1 Conventional whole-cell recording

Under voltage clamp, the current through the preparation is measured by means of a feedback system that maintains a set potential difference between the two electrodes (pipette and bath electrode) by injecting current. The injected current is the opposite of the current generated by the preparation. Thus, if the two electrodes are separated by a membrane, the injected current required to maintain the set potential directly reflects the membrane current (see also Chapter 2). Most patch clamp experiments involve recording the summated activity of the whole cell membrane, whereas in more detailed studies the activity of single ion channels is recorded. For whole-cell recording, the total cell membrane must be positioned between the two electrodes, which implies that the pipette electrode must have a low-resistance connection with the intracellular side of the cell. After gigaseal formation, this situation can be obtained by breaking the patch of membrane under the pipette tip while leaving the seal resistance intact. The pipette solution is then in direct contact with the cytoplasm, and mixing of solutions takes place. Because the volume of the pipette solution is many times more than the cytoplasm, if given time the cytoplasm will be replaced by the pipette solution. In most whole-cell experiments the pipette solution therefore largely resembles the ionic composition of the cytoplasm (see Table 4.1 for example). After gigaseal formation, but before a breakthrough is attempted, it is helpful to switch the pipette potential to a value near the membrane potential. This will make the patch more stable and prevents a sudden depolarisation of the cell when a breakthrough is achieved. A successful breakthrough is clearly indicated by the current response to a test pulse, because a relatively large capacitive transient should appear (Figure 4.6).

The transient is formed by the cell capacitance in series with the pipette plus the access resisitance (which make up the series resistance, see Section 2.3.5). A good cell capacitance in combination with a low series resistance results in a large, quick transient. In addition, there will be a slight increase in steady-state current, reflecting the membrane resistance in series with the series resistance (Figure 4.7).

4.2 WHOLE-CELL MODES

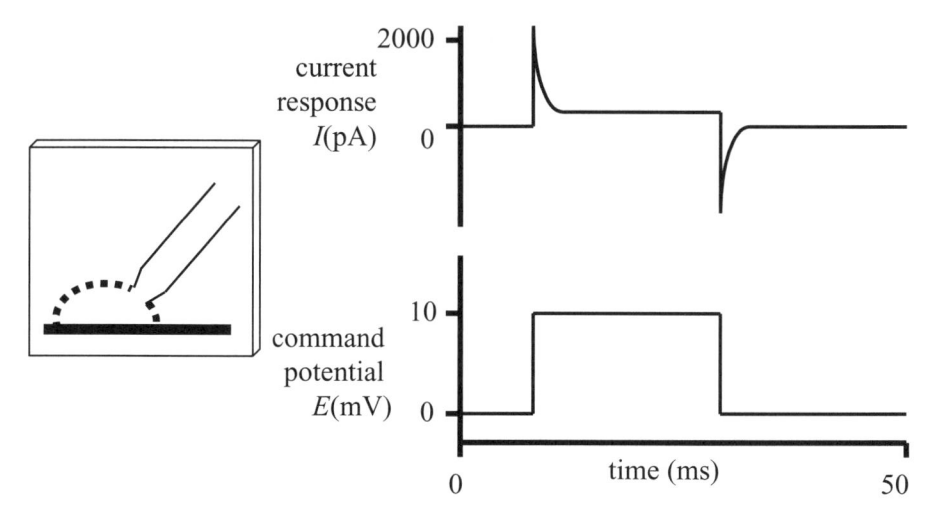

Figure 4.6 A typical whole-cell current response A positive or a negative test pulse can be used, whichever has the smallest likelihood of activating ion channels

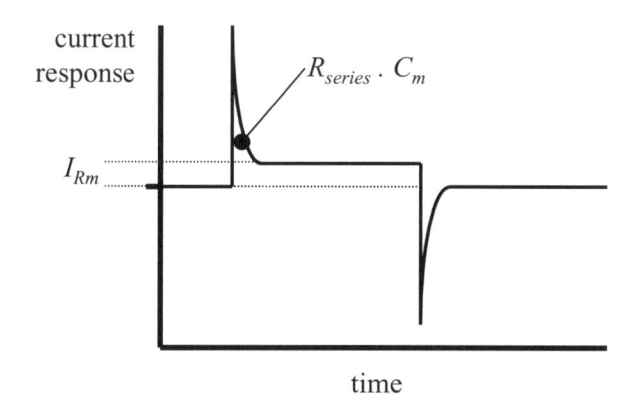

Figure 4.7 A whole-cell current response to a square test pulse is made up of a transient RC response formed by the series resistance $R_{series} (= R_{pipette} + R_{access})$ and the cell capacitance C_m, and an ohmic or steady-state component determined by the membrane resistance R_m

There are two methods to break the patch of membrane under the pipette.

1. A suction pulse can be applied through the pipette pressure tubing. The suction is applied by mouth. This is a skill that needs practice. The

amount of suction and whether or not it is applied steadily or in a pulse depends on the pipette size, cell type and the particular set-up. A low-resistance pipette (about 2 MΩ or less) covers a relatively large patch that usually can be disrupted with a gentle kiss. With smaller pipettes a larger suction pulse or a maintained increasing suction is required, although the suction must be removed as soon as the patch breaks. In the case of very small pipettes (>7 MΩ) suction might not work at all and another method might be called for.

2. A large current pulse can be sent through the pipette. Some patch clamp amplifiers have a 'zap' function that does just that. Zapping works best with small pipette tips because the current is concentrated on a small patch. It is prudent to start with a short pulse duration (or pulse amplitude, whatever can be set) and increase the pulse successively until a result is obtained. This prevents the cell from being blown away by too large a current.

Although each situation is different, a general strategy with medium-size pipettes (2–7 MΩ) might follow these steps until the whole-cell configuration is reached:

- Suction pulses without excessive force.
- 'Mild' zapping (without damage to the seal resistance, as indicated by increased noise levels and instability).
- A zap immediately followed by a suction pulse.
- Increased zap pulse size and suction force.
- Continuous suction increasing in force.
- Incrementing zap pulses while strong suction is applied until the patch breaks, the seal breaks, the amplifier starts to smoke or your fillings come loose!

A common source of problems using suction pulses is that, if the pipette is not tightly fixed to the pipette holder, the pipette will move when suction is applied, disrupting the patch. It is therefore a good idea to glance at the cell and pipette occasionally, instead of the current response, while applying suction pulses to make sure that the position of the pipette is stable.

Once the typical whole-cell current response appears, the membrane potential can be measured by quickly switching to current clamp (with zero current). The membrane potential is an important parameter to record because it is a criterion for the quality of the configuration and, if averaged over a number of observations, can be used in cell-attached patch recording analysis (see Section 4.3.2). The capacitive transient in the whole-cell current response can usually be cancelled by setting two knobs calibrated for series resistance and membrane capacitance. Membrane capacitance provides a rough estimate of membrane surface area (see Section 2.2.3), assuming that the average square centimetre has a capacity of about 1 µF.

Quality control

For a recording to be included in the analysis it must meet the quality criteria, which should be drawn up beforehand. The details are up to the experimenter, but a typical list of criteria for a whole-cell recording could look like this:

1. *The starting seal resistance must be better than 1 GΩ.*

2. *The series resistance must be lower than 20 MΩ and stay that way throughout the recording.* A high series resistance (due to a high access resistance because the pipette resistance does not change) is undesirable because voltage clamp of the cell membrane is adversely affected (see also Section 5.2.3). The time constant τ of the capacitive transient in the current response is proportional to the series resistance, so a slow capacitive transient is a bad sign. A frequently occurring problem is gradual resealing of the patch after breakthrough, apparent through spreading of the capacitive transient. Sometimes gentle manipulation (suction, pressure or zapping) can recover the situation, but it is likely that the problem will need attention throughout the experiment. Understandably, it is less likely to occur with larger pipettes, so decreasing the pipette resistance is often a successful remedy.

3. *The membrane potential must be more negative than −50 mV if normal high-potassium intracellular solution is used.* The ionic composition of the intracellular medium changes as soon as the patch is disrupted, so subsequent measurements cannot be compared directly.

4. *Cell capacitance and resistance must be stable.* If the membrane

becomes leaky or the seal resistance breaks up, the ohmic component of the current response will increase, as in Figure 4.5 (right centre panel). A leaky membrane and a breaking seal can sometimes be distinguished by the capacitive transient, which can remain when the cell becomes leaky but disappears when the seal breaks up.

4.2.2 Perforated patch recording

It was mentioned in the previous Section that, as a consequence of disrupting the patch of membrane under the pipette tip, the soluble components of the cytoplasm are replaced by the pipette fluid. This provides the experimenter with control of the intracellular medium, although manipulation of this medium *during* an experiment is not straightforward (see Section 3.2.4). However, the replacement of the cytoplasm by pipette solution represents an experimental manipulation for which there seems to be no proper control experiment for comparison; this is a scientifically unsound situation. In addition, there are many situations where replacement of the cytoplasm is undesirable, primarily when the subject of study involves intracellular signalling. Classic whole-cell recording could dilute or wash out crucial elements in the signalling cascade. With the invention of perforated patch recording, a compromise between good electrical access and preservation of the intracellular milieu has become available.

The principle of perforated patch clamping relies on the action of drugs used as antifungals for many years. These drugs create holes in membranes of a very distinct size, permeable to ions but not larger molecules, most critically second messenger molecules such as cAMP. Initially nystatin was used, but more stable results are obtained with amphotericin B. The key to successful perforated patch recording is controlled application of the drug. If the drug is applied before a gigaseal is formed, then the membrane will be damaged in the area where a tight seal should form, and no good seal can be achieved. A delay must therefore be built in to ensure that the perforating drug is only exposed to the patch of membrane under the pipette after a seal is formed. Although a pipette perfusion system could be used (see Section 3.2.4), it is much more practical to fill the pipette tip with normal intracellular solution and fill the remainder of the pipette with intracellular solution containing perforating drug. Diffusion then will gradually introduce the drug to the patch and the development of the whole-cell current response can be monitored.

4.2 WHOLE-CELL MODES

Procedure

The antifungals must be dissolved in ethanol or dimethylsulphoxide and added to the intracellular solution. The pipette tip can be filled with normal intracellular solution using suction from a large syringe connected to the back of the pipette with some tubing. The pipette then can be filled with intracellular solution containing perforating agent. Typical concentrations for nystatin or amphotericin B are 5–20 $\mu g\,ml^{-1}$ from 1000-fold stock. (Note that the intracellular solution does not require ATP or GTP.) Gigaseal formation is performed as usual, but then the patch is left intact and the current response to a test pulse is monitored. If all is well, the cell capacitance should manifest itself after a few minutes, with decreasing access resistance as indicated by a progressively faster capacitive transient. After perhaps 10 min the capacitive transient is stable and the experiment can begin (Figure 4.8).

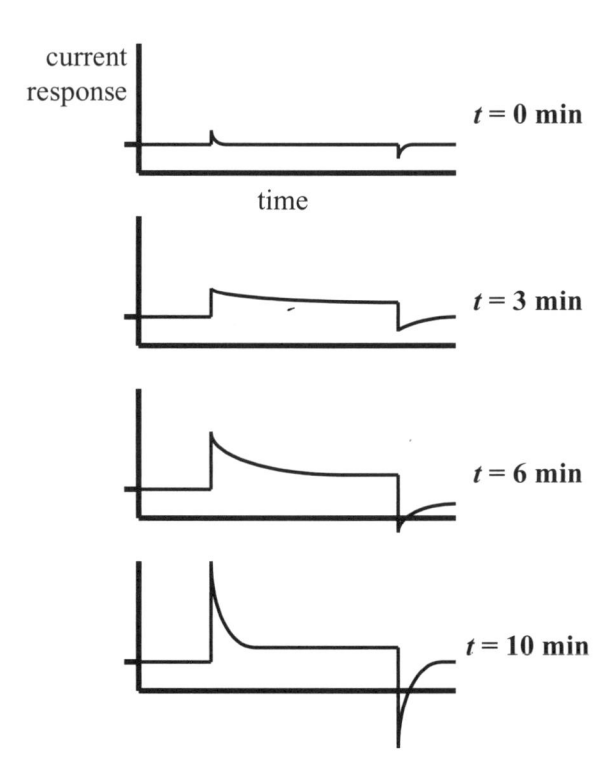

Figure 4.8 Establishment of a perforated patch configuration. The gigaseal is formed at $t = 0$ min. The patch initially appears to become leaky (as if the seal breaks up, but without the increase in noise) and then the capacitive transient develops

The critical factor in perforated patch recording is timing. The amount of normal intracellular solution in the pipette tip determines the time available to obtain a gigaseal before the perforating agent reaches the membrane. The clock starts ticking the moment the pipette is back-filled with intracellular solution containing antifungal. If there is excess normal intracellular solution in the tip, it can take a very long time for the drug to take effect, which may be incomplete as well. As with many details in patch clamping, every situation is slightly different and intelligent trial-and-error is the only way to establish the details for yourself. Once a good whole-cell current response is obtained, it is likely to be stable (particularly with amphotericin B) but it is good practice to make regular checks (see Ueno et al., 1992, Yawo and Chuhma, 1993, and Spruston and Johnston, 1992, for further reading).

4.3 Single-channel Modes

4.3.1 General notes

The patch clamp technique allows the study of single proteins in action, which is a feat of stupendous nanotechnology established before the word was even invented. There are three configurations in which single channels can be studied: cell-attached patch and two excised patch configurations. The technical implications of the configurations are discussed in Sections 2.3.5 and 2.3.7. The choice of single-channel mode is dependent on the hypothesis to be tested, because each mode allows easy manipulation of different media or, in the case of a cell-attached patch, the cell to which the patch belongs. This Section describes how to obtain single-channel modes in practice.

General differences between setting up single-channel recording and whole-cell recording are mainly a consequence of the 100–1000-fold increase in sensitivity of recording. Single-channel data almost always have a small signal-to-noise ratio, so noise control is very important. For example, the quality of the gigaseal must be at least an order of magnitude better than in whole-cell work, and external noise must be undetectable, i.e. lower than the intrinsic noise. By the same token, the intrinsic noise levels must be minimised, e.g. by reducing stray capacitance and using low-noise pipette holders. Some patch clamp amplifiers even have separate ultralow-noise circuits for single-channel recording.

Single-channel activity is quite different in nature from whole-cell data. The opening and closing of channels is a stochastic and practically instan-

4.3 SINGLE-CHANNEL MODES

taneous process caused by the protein switching between conformations. This behaviour is reflected in recordings as instantaneous jumps between current levels (Figure 4.9). The jumps are so quick that the rate of rise or fall during jumps as recorded is usually determined by the filter settings of the amplifier. Consequently, recording and analysis of single-channel data are very different from working with whole-cell data (see Chapters 5 and 6).

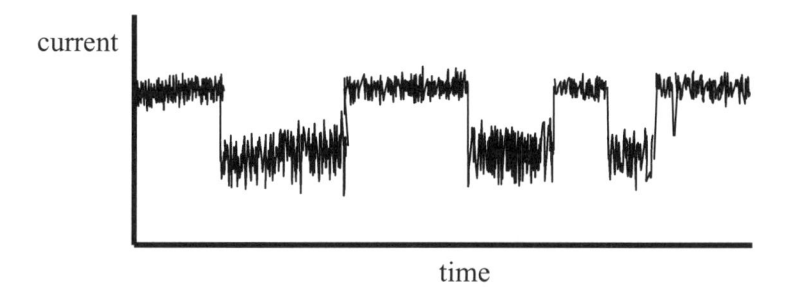

Figure 4.9 An example of single-channel activity as it appears on the oscilloscope or monitor. The current level jumps between two average values as the channel opens (lower level) or closes (upper level). The incomplete-looking transition at the end of the trace is actually complete but not captured by the apparatus owing to the low-pass filter. The current scale is of the order of pico-amperes, whereas the time scale can be of the order of milliseconds to seconds, depending on the channel type

As will become very apparent when analysing single-channel data, the ideal single-channel recording contains the activity of only one channel. There are measures that the experimenter can take to increase the chance of this occurring:

- Use of channel blockers and/or choice of physiological salt solution to block or mask activity from channels other than those under investigation.

- Choice of pipette tip diameter with the largest probability of capturing one channel.

The task is not simple and in some cases is impossible because certain channel types are known to associate in clusters. Even if only jumps between two current levels are seen, what is the guarantee that the transitions are caused by activity from only one channel? (see also Chapter 6).

4.3.2 Cell-attached patch

In all single-channel modes, the tiny patch of membrane under the pipette is the subject of study. It follows that, once a gigaseal is formed, a single-channel recording situation is established without any need for further manipulation. The cell-attached patch configuration is very desirable in several ways:

- The patch is *in situ*, so is expected to behave 'physiologically', e.g. with respect to structural integrity and intracellular modulation.
- There is a high success rate in obtaining the configuration.
- There is little deterioration in the configuration over time.

The major disadvantages of the cell-attached patch configuration are twofold:

- The media on either side of the patch can be manipulated only with great difficulty: the cytosolic side can be influenced indirectly, e.g. by evoking a second messenger signal in the cell, whereas the extracellular side can be changed only *during* an experiment by using the same trick as in perforated patch recording, or by pipette perfusion, which is practically impossible in single-channel mode.

- The membrane potential of the cell is intact and unknown, as is the exact ionic composition of the cytoplasm. As a consequence, precise analyses of reversal potentials and related parameters (relative permeabilities, etc.) are very difficult to carry out in this mode. The problem can be aggravated by an active membrane potential that changes during the experiment. However, if the membrane potential is known to be stable the previous whole-cell studies might have provided the experimenter with some idea of the average and range of membrane potentials for a number of cells. Provided that the range is not too large, the average value can be assumed in cell-attached patch recordings. Another method for gaining some idea of the membrane potential would be to break the patch after the experiment, although the measurement would have to be very quick as in cell-attached patch recordings the pipette solution will usually be extracellular solution. You would also have to be sure that the membrane potential is unchanged by, or has recovered from your experimental procedures. It is obvious that both of these methods are inexact.

4.3.3 Excised patches

Single-channel studies using excised patches provide ultimate control for the experimenter. In outside-out patches the cytosolic side faces into the pipette, whereas for inside-out patches the extracellular side faces the pipette. Because the bath solution can be changed easily during the experiment and the pipette solution cannot, the choice of excised patch configuration is usually straightforward and depends on which part of the channel protein is to be modulated. Voltage clamp of the patch in both situations is very good because the patch is the only (very high) resistance between the two electrodes. In practice, outside-out patches are easier to work with than inside-out patches for two reasons: it is easier to obtain the outside-out configuration, and in the inside-out configuration the bath solution must be replaced with intracellular solution.

Outside-out patch

The first part of the procedure to obtain an outside-out path is exactly the same as for classic whole-cell recording, including filling the pipette with intracellular solution and applying a holding potential at a realistic membrane potential value. Once the whole-cell configuration is established, the pipette is gently pulled away from the cell. In some cases it is best for the stability of the patch that the amplifier is switched to current clamp before this manoeuvre is performed. The membrane will break around the pipette and reseal to form a new patch (see also Figure 2.23). This can be observed as a 'resealing' event on the current response to the test pulse (Figure 4.10).

If the seal has been compromised during the process, the final resistance will be less than 1 GΩ and the noise level will be high. Resistance and noise therefore provide instant controls to check the quality of the patch. In the outside-out excised patch configuration there is no need to change the bathing solution and at this point the experiment can start.

Inside-out patch

The objective of using inside-out excised patches is usually to expose the intracellular side of the patch to different conditions during a recording. The easiest way to change a solution in a set-up during an experiment is to perfuse the bathing chamber, hence the patch under study must be

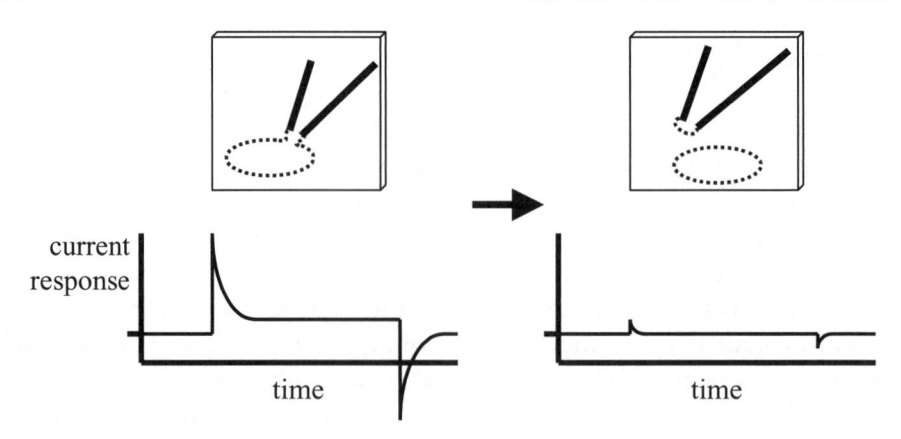

Figure 4.10 The current response to a test pulse shows resealing when the pipette is pulled from a cell. The integrity of the seal is immediately obvious

configured so that the cytosolic side faces the bath. Once a gigaseal is formed, the pipette is pulled away without breaking the patch under the tip. Similar to the outside-out patch procedure, the bond between the membrane and the pipette glass is stronger than the structural strength of the membrane, and the membrane will tear around the pipette. However, the subsequent resealing of that membrane is now an awkward complication, because a vesicle is formed (see Figure 2.23). The vesicle cannot be 'seen' electrically; there should be no change in the observed gigaseal. The outward-facing membrane can be destroyed by briefly lifting the pipette tip out of the bathing solution. This takes practice and patience. The procedure is complicated by the fact that there is no quick way of telling if the exposure was too short or not (although you will know straight away if it was too long). The added complication is that the bathing solution now faces the cytosolic side of the patch, so in most experiments it must be replaced with intracellular solution. This takes time and might not go down well with the other cells in the bath. 'Disagreeing' cells tend to deteriorate and die. If the cells are rare (carefully grown cultures, freshly dissociated cells or brain slices), these experiments can be costly and time-consuming. A solution to this problem might be local perfusion, as explained in Section 3.2.4.

5
Whole-cell Protocols and Data Analysis

5.1 Standard Cellular Parameters

It is good practice to keep a record of standard parameters of each cell being patched. The list that should appear in the laboratory logbook for each cell could include:

- $R_{pipette}$: pipette resistance, recorded with the pipette in the bath but not on the cell.

- R_{seal}: seal resistance, recorded after gigaseal formation.

After the whole-cell capacitive transient is established (see Section 4.2):

- E_m: membrane potential, measured by briefly switching to current clamp mode ($I = 0$) as soon as the patch is disrupted.

- R_{series} and C_m: series resistance (= access resistance + pipette resistance) and cell capacitance, respectively, recorded after capacitive transient cancellation.

- R_m and R_{leak}: the parallel membrane resistance and leak, respectively, measured from the ohmic component of the current response to a test pulse. The magnitude and sign of the resting current provides a hint of the proportions between the two resistances. The leak resistance is not ion selective, so the leak current has a reversal potential of 0 mV. The resting membrane conductance (reciprocal of resistance) at rest usually favours potassium, so the reversal potential for this conductance will be near E_K if a normal high-potassium intracellular solution is used (see Section 2.2.1). It follows that if the leak resistance is dominant (low), there will be an inward resting current at about −60 mV, whereas if the resting cell conductance is dominant, the resting current will be near zero or even positive (Figure 5.1).

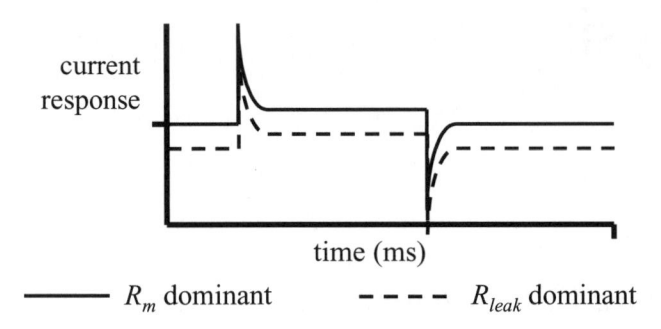

Figure 5.1 At a holding potential of −60 mV or even more negative, a resting current around zero is indicative of a high leak resistance

- per cent R_{series} compensation, if used (see Section 3.4.2).

These parameters are important quality controls, as described in Section 4.2.1, and can also help in identifying cell types in heterogeneous cell populations.

5.2 Voltage-activated Currents

The ensemble of voltage-activated channels that a cell possesses is an important signature of the cell type and very easy to determine. Even if the conductance under study is not one of the voltage-activated ones, it is a good idea to record this signature if possible at the start of an experiment and add it to the standard cellular parameters recorded for each cell.

5.2.1 Introduction to pulse protocols

Voltage-activated currents are evoked by stepwise changes in holding potential, similar to the test pulse used to monitor the status of the patch.

5.2 VOLTAGE-ACTIVATED CURRENTS

The commonest protocol to record voltage-activated currents is to apply a series of voltage steps of fixed length but varying amplitude. The steps are spaced wide enough to consider the current responses to them as independent. The voltage steps usually originate from the holding potential, which is chosen near the resting potential for stability and because most voltage-activated ion channels are closed at this potential. The exact parameters of the pulse protocol depend on the cell and the experimental conditions. Most patch clamp software allows the programming of very complex pulse protocols but it is likely that a simple pulse protocol, as shown in Figure 5.2, is the one you will use the most. The pulses and the current responses are shown individually for clarity, but it is much more common to present these data in one combined graph showing an ensemble of currents. This is demonstrated in Figure 5.3. By convention, currents caused by positive charge flowing *to* or negative charge flowing *from* the ground electrode are shown deflecting upward. In the case of whole-cell or perforated patch currents, these currents are outward, i.e. from the cytosolic side of the membrane to the extracellular side. Note that this convention says nothing about the direction of actual ion flow: if the ions carrying outward currents are positive, the ions leave the cell; if the ions are negative, they enter the cell! The convention is not always strictly adhered to (particularly in single-channel modes, where the situation can get even more confusing), so careful reading of the literature is warranted.

Passive components

There are a great number of important observations to be made on the example experiment depicted in Figures 5.2 and 5.3. Firstly, the pulse protocol starts at −60 mV (the holding potential) and steps to more positive values in increments of 20 mV. Each step has a duration of 50 ms, and the time between the start of each step, although not depicted, is 1 s. The current response at rest (before the pulse starts) is about zero. The smaller depolarising steps elicit a current response that is almost instantaneous, with only a small transient at the start and the end of the pulse. Both the transients and the ohmic response (sometimes called steady-state or plateau response) are proportional in size to the test pulse. The transient at the end of the current response is the exact inverse of the transient at the start. It seems that these elements in the current traces behave passively in response to the voltage steps. The transients in fact are remnants of the capacitive transients that were cancelled in the beginning of the experiment. Cancellation, however, is never perfect because the real capacitances

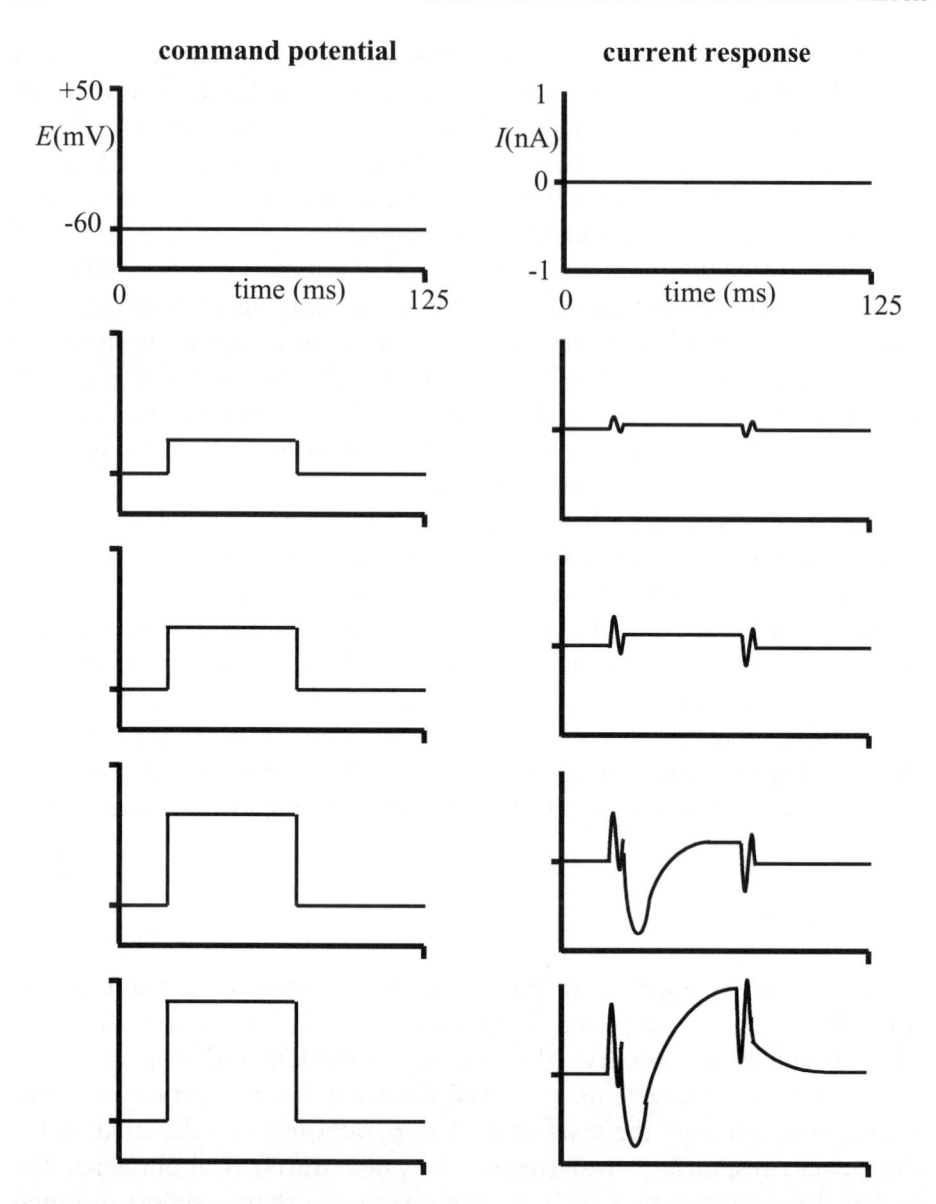

Figure 5.2 Example of common current responses, as seen in neurones to incrementing depolarising voltage steps. The voltage step length is 50 ms, the increment is 20 mV per step, time between the starts of each step is 1 s and the holding potential is −60 mV

and resistances in the system do not behave like an ideal RC circuit, and the imperfections are magnified by larger voltage steps. The ohmic element is caused by a combination of the leak resistance R_{leak} and the cell's background conductance R_m (as recorded at the start of the experiment).

5.2 VOLTAGE-ACTIVATED CURRENTS

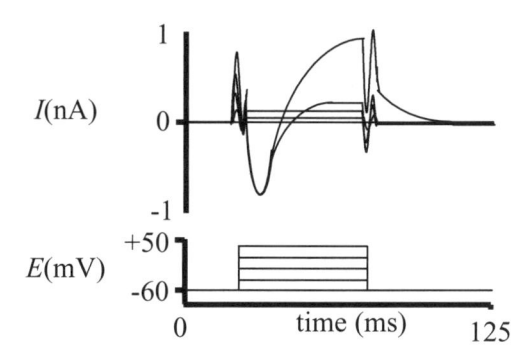

Figure 5.3 Common method of representation of data from Figure 5.2

Passive components can be eliminated in data presentations by using positive/negative subtraction (see Section 5.2.2).

Active components

At a depolarisation of 60 mV to a membrane potential of 0 mV, a large transient inward (downward) current is seen. The current is clearly activated by voltage because its amplitude is not proportional to the voltage step size and it does not appear symmetrically in the depolarisation and repolarisation phases of the step pulse. Another interesting feature is its brevity, which indicates that it inactivates over time. Further depolarisation superimposes a second, slower voltage-activated current, this time outward, onto the current response. The outward current does not seem to inactivate within the duration of the depolarising pulse, and leaves a small inactivating outward current after the membrane potential has returned to its resting holding value. These currents represent prototypes of very common ion currents. The inward current could be mediated by excitatory sodium or calcium channels, whereas the outward current looks like a potassium current. Methods to verify this and study the currents in detail will be discussed in the following Sections. Firstly, some notes on signal conditioning are presented.

5.2.2 Signal conditioning and positive/negative subtraction

If the cell type under investigation is new to the experimenter, then great caution should be taken that a thorough inventory is made of the voltage-dependent currents. The amplification factor is self-evident and should be chosen so that the maximum analogue-to-digital (AD) conversion resolution is achieved without clipping (see Section 3.4.4). It is worthwhile playing with a range of time scales. For example, it is not guaranteed that time-dependent inactivation of the outward current in Figure 5.3 is absent, it might only show up in longer depolarisations.

Filter and AD conversion settings should be adapted to the signal. If a good overview of the voltage-dependent currents is visible in traces of 150 ms and a trace is made from 1024 or 2048 data points, then the sampling rate for AD conversion would be 6.8 or 13.6 kHz. As a rule of thumb, the low-pass filter would have a cut-off frequency of not higher than half the sampling rate to avoid aliasing (see Section 3.4.4), so the setting would be 3.4 or 6.8 kHz.

The situation can be complicated by the co-occurrence of several voltage-dependent currents with very different kinetics. In this case, recordings can be made on two different time scales. This can be done in separate experiments, but some software packages allow programming of different sampling rates in one pulse protocol, so one experiment can cover different time scales (Figure 5.4). Note that if the filter setting cannot be changed together with the sampling rate in such experiments, it might be necessary to add filtering off-line for the slower sampling rate.

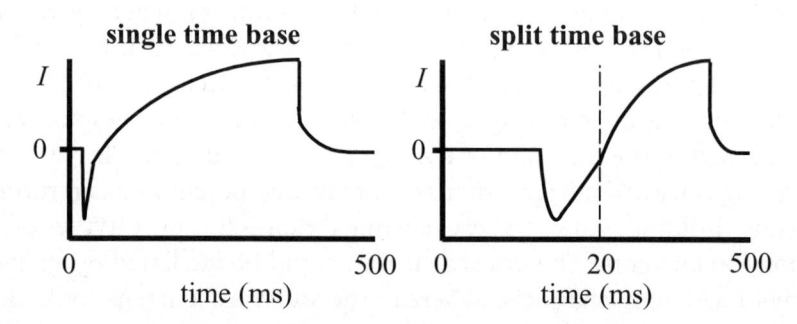

Figure 5.4 Dual time-base recording allows detailed recording of consecutive fast and slow events, or *vice versa*, in one experiment

Positive/negative subtraction

Passive components can be removed from pulse protocol data by a relatively simple but very effective method called positive/negative (P/N)

5.2 VOLTAGE-ACTIVATED CURRENTS

subtraction. It is applied very widely and is built into most patch clamp software. The condition that must be met for P/N subtraction to be valid is that there must be a range of membrane potentials for the cell under investigation where no active currents are present, usually around the resting membrane potential. A voltage pulse within that range will evoke a current response with no active components, just remnants of capacitive transients and an ohmic response made up of R_{leak} and R_m. Because the passive components are proportional to the size of the voltage pulse, these components can be scaled to those generated by any test pulse, including pulses that also evoke active components. The passive response can then be subtracted from the response that includes active and passive components, to leave a 'clean' trace of active currents. An example of the process is shown in Figure 5.5.

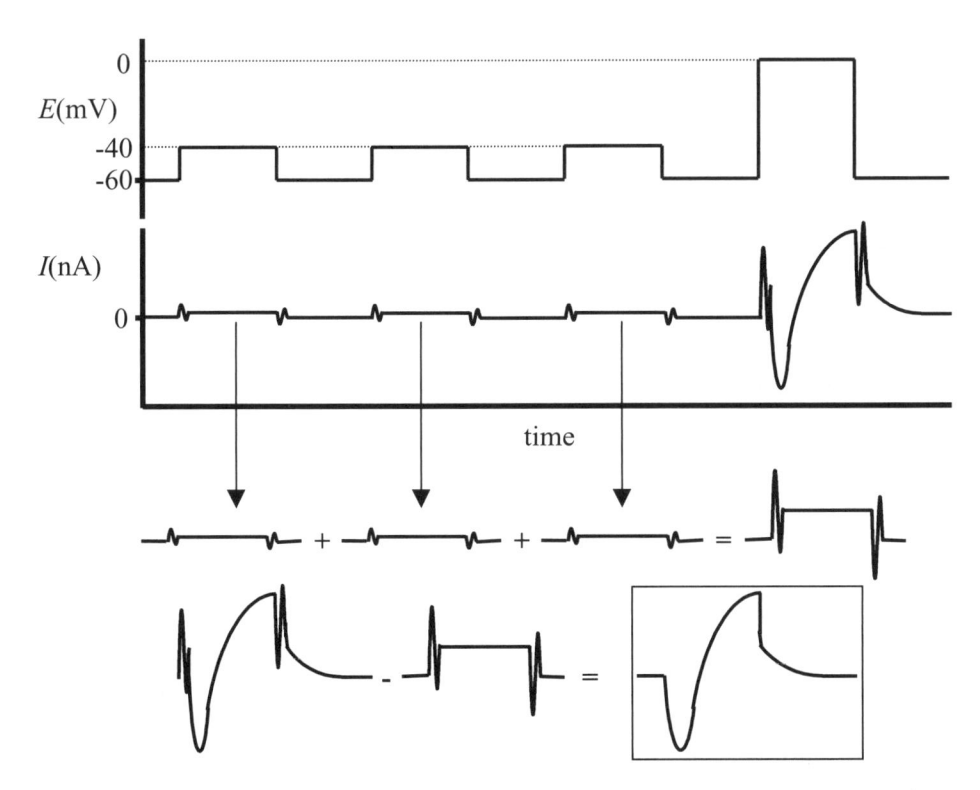

Figure 5.5 Example of a P/N subtraction. Three current responses to a 20 mV voltage pulse are summated and then subtracted from the current response to a 60 mV voltage pulse

Some important considerations in using P/N subtraction are listed below:

- In the example of Figure 5.5 three passive component responses (sub-episodes) are added instead of mathematically scaling a single response with a factor of three. The advantage of this is a reduction in noise in the culminated response. The noise level decreases with increased number of sub-episodes. The disadvantage of a large number of sub-episodes is that the experiment takes longer and slow changes in the patch become significant.

- Application of P/N subtraction leads to loss of information, most importantly of resting current. Some workers argue that P/N-corrected traces do not properly represent raw data. It is even possible, if monitoring only corrected traces during an experiment, to work with a severely deteriorated cell without realising it, and wonder why the active currents are so small and/or distorted. If your software package records only the corrected trace, then it is good practice to record an additional ensemble without P/N or to apply P/N off-line. The latter is possible by scaling a current response with only passive components and to use that to subtract from traces with active currents. The noise in such responses is, of course, not as low as from summated separate sub-episodes.

- The time between sub-episodes must be long enough for the membrane to recover fully.

- Most mishaps using P/N subtraction occur when it is applied to more complicated pulse protocols without due consideration to the condition that no voltage-activated current must be evoked in recording the passive components.

- The experimenter must be aware that passive components can change during the experiment. As mentioned earlier, the leak can change. The passive components also include the resting membrane resistance, which, among others, usually includes the resting potassium conductance. This is an ion-channel-mediated conductance and therefore is prone to modulation. If resting conductances are influenced by the experiment, then the use of P/N subtraction is inappropriate.

5.2.3 Space clamp artefacts

In the discussion of voltage clamp experiments it is assumed so far that the model presented in Figure 2.7 is valid, i.e. the intracellular and extracellular media are represented as conductors separated by a cell membrane. The media are not quite as good in conducting current as metals, partly due to the fact that current is carried by ions instead of electrons. In round, non-branching cells the intracellular medium can, for the purpose of voltage clamping, be considered a full conductor. The voltage is clamped over the significant resistance in the circuit i.e. the cell membrane (see Section 2.3.3). In branched cells such as neurones *in situ*, voltage clamp of the membrane is impaired by significant voltage drops over volumes of intracellular medium between the pipette electrode and the bath electrode. This is illustrated easiest by a graph incorporating resistances. In a round cell, the intracellular resistance is much lower than the membrane resistance for every part of the cell membrane (Figure 5.6). In a dendrite or axon, the thin, long volumes of cytoplasm can be considered as resistances in series, which accumulate with length of the branch. The cytoplasmic resistance now becomes significant in relation to the membrane resistance

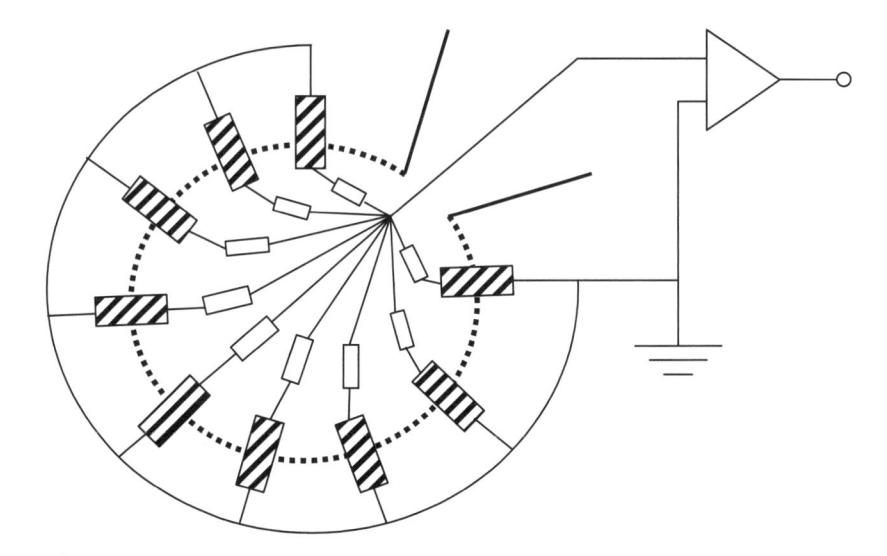

Figure 5.6 A round cell has negligible intracellular resistance (white) compared with membrane resistance (hatched) in all directions, aiding good voltage clamp of the membrane

and, as a result, voltage clamp of areas of membrane distant from the pipette electrode will be poorly clamped.

A second distorting factor is the cell capacitance along the membrane lining the branches. The effect of this capacitance is to delay changes in clamp voltage (see Section 2.2.3). If we add capacitance to the branch circuit from Figure 5.7, we obtain the standard model to illustrate cable properties (Figure 5.8). It is known from this discipline in electronics that the build-up of resistance is proportional to the length of the stretch of cytoplasm that needs to be traversed by the current, and is inversely proportional to its diameter. It follows that long, thin branches are most susceptible. The phenomenon of poor voltage clamp due to significant

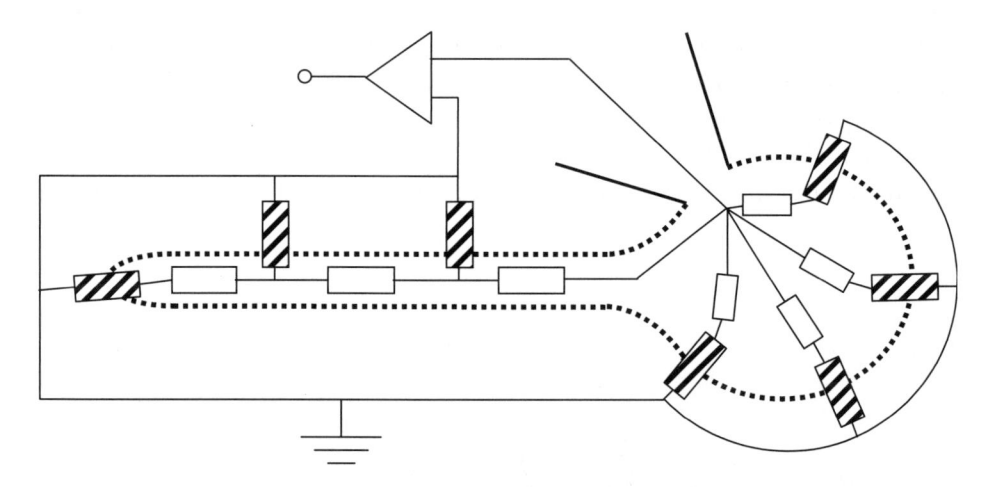

Figure 5.7 In an elongated or branched cell the resistance between the pipette electrode and some parts of the cytoplasm becomes significant in relation to the membrane resistance (hatched), leading to poor voltage clamp of the latter

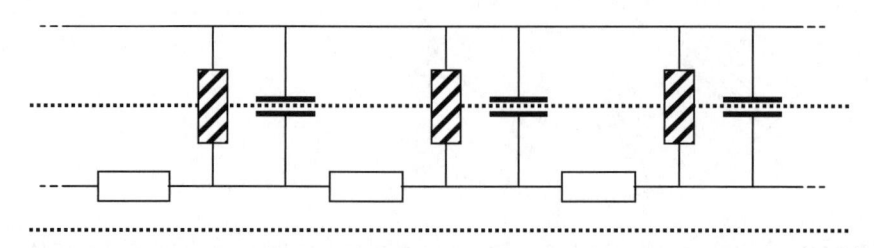

Figure 5.8 The standard model to describe cable properties. It can explain, for example, the spatial exponential decay of a current injection along the length of an axon

5.2 VOLTAGE-ACTIVATED CURRENTS

cytoplasmic resistance is known as space clamp and is recognisable in pulse protocol experiments as sluggish, incomplete current responses (Figure 5.9). Although cable properties are not relevant to patch clamping, except in space clamp, and therefore are discussed here no further, their consideration is very useful in describing action potential propagation and how neurones process synaptic potentials (see also Spruston et al., 1993).

The space clamp condition is an important restriction in the preparation (cell) type that is suitable for voltage clamp experiments. Generally, unsuitable cells are branched neurones *in situ* (in slices), or neurones grown for a considerable time in culture, when they have developed into a network. Patch clamping can still be applied in these cells in single-channel modes, or in whole-cell mode if it is known that the ion channels under study are restricted to the soma. Current clamp (membrane potential recording) is not subject to space clamp and is an important tool in these cells, e.g. for recording of synaptic potentials. Some ionic conductances are more prone to space clamp distortion than others. Because space clamp results in unreliable membrane voltage measurement, currents through channels with steep voltage dependencies in activation and/or inactivation (see Section 5.2.7) will be most affected. An example of such a current is the neuronal tetrodotoxin (TTX)-sensitive sodium current, which looks somewhat like the current depicted in Figure 5.9. An important indication of space clamp is provided by the property that distortions such as sluggishness of the current are greatest in the activation range of the current, so that an ensemble of sodium currents will show maxima at different time points, similar to the traces in Figure 5.9. The spread in these time points is therefore a rough indication of the amount of space clamp in the cell.

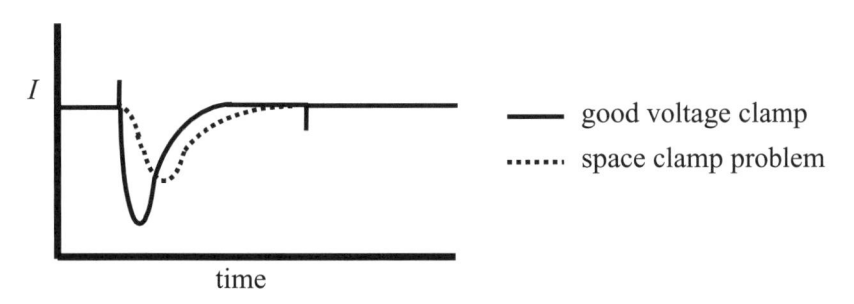

Figure 5.9 Example of space clamp effect. Currents are distorted and incomplete

5.2.4 Isolation of a homogeneous population of channels

All cells possess a multitude of channels and it is likely that a voltage step protocol will activate several populations. The isolation of a single population to study in detail is not always easy. The following lists some common methods:

- *Shape of the current response.* An obvious sign of a multi-population response is two or more phases in the response that clearly have different voltage dependencies and kinetics. An example is shown in Figure 5.3, where an inward current appears at a certain voltage followed by an outward current at another, more depolarising voltage.

- *Drugs.* There are many pharmacological tools available that effectively block the known ion channels. Some of these are very specific natural toxins, such as the aforementioned TTX, whereas others knock out most conductances to certain ions. The latter are most effective in excluding the very large groups of calcium channels (block by 1–2 mM of cadmium ions in the extracellular solution) or potassium channels (block by replacing intracellular potassium by caesium or using a 10–30 mM concentration of tetraethylammonium in the extracellular or intracellular solution).

- *Manipulation of ion concentrations.* Changing ion concentrations in intra- and extracellular media can affect ion conductances differentially, depending on their ion specificity.

The trouble is that when studying a newly discovered current none of these methods positively secures that the current is mediated by one type of channel, although they can all contribute to that notion. In practice a single population is identified by simple activation and inactivation kinetics (see Section 5.2.7), and if possible specific sensitivity to a drug. The population is considered single until a manipulation reveals otherwise. There is, however, one direct way of knowing that a current is carried by one type of ion channel: if single-channel data can be summated to form a pseudo-whole-cell current, then true whole-cell currents that match its kinetics can reasonably be assumed to be mediated by that type of channel only (see Section 6.3.3). This method is of course only applicable if single-channel studies are a part of the investigation, which is often not the case.

5.2.5 Current–voltage relationships and reversal potential

The relation between membrane voltage and current size provides many of the basic characteristics of a voltage-activated conductance. These include, for example, the voltage range of activation and equilibrium potential. The latter is determined by the conductance's ion specificity, hence an easy indication of the species of ions permeant to an ion channel is the current's reversal potential, as discussed in Section 2.2.1. If the permeability of the ion channels under study is strongly dominated by one ion species, then the reversal potential will be very near that ion's equilibrium potential. If more ion types can pass, the reversal potential will be a function of the relative permeabilities and equilibrium potentials of those ions. To determine the reversal potential of a current in practice, all that is required is to create a current–voltage (I/V) relationship from an ensemble of currents obtained using a step protocol, as discussed in Section 5.2.1. The easiest situation occurs when all ion channels are open in the voltage range around the reversal potential, and reversal can be observed directly. An example of this is the TTX-sensitive, voltage-dependent sodium current, of which the result of a step protocol experiment and the I/V relation derived from it is shown in Figure 5.10.

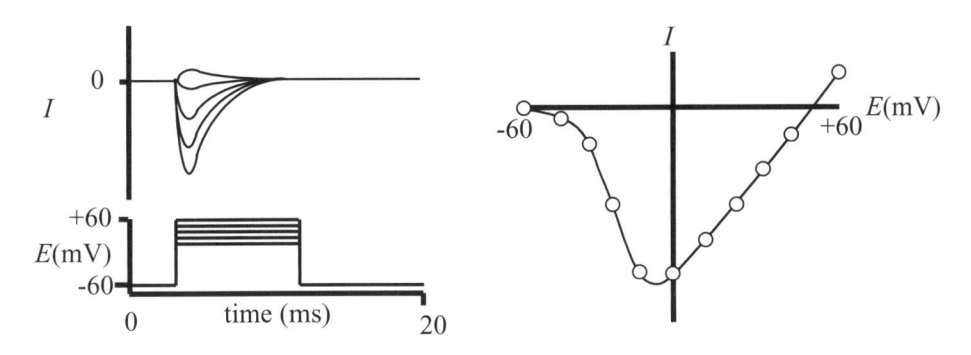

Figure 5.10 Results of a step protocol in the study of an inward current. Depolarising voltage steps with increments of 10 mV are applied from −60 to +60 mV. In the left-hand panel only raw data from the last five steps are shown for clarity. The right-hand panel shows the relation between peak-evoked current and step potential. The maximum peak-evoked current can be of the order of several nano-amperes in large neurones

Most of the channels are closed at potentials negative from −40 mV but open when the potential is stepped to more positive values. Steps to potentials more positive than about 0 mV will open practically all channels. At such potentials the conductance is constant and current size varies only with the driving force for the permeant ions (see Section 2.2.1) in a linear relationship. The relation between peak inward current and membrane potential shown in Figure 5.10 has a linear part between 0 and 60 mV, indicating that a constant number of channels are open. The reversal potential for sodium at room temperature with $[Na^+]_{out} = 120$ mM and $[Na^+]_{in} = 6$ mM is $+50$ mV (see Section 2.1.2). The recorded inward current becomes outward around such a potential value, strongly suggesting that it is carried by sodium.

The maximum conductance of a population of ion channels can be determined easily by calculating the maximum slope between current and voltage with the reversal potential as origin (Figure 5.11). This line usually coincides with the linear part of the I/V relation.

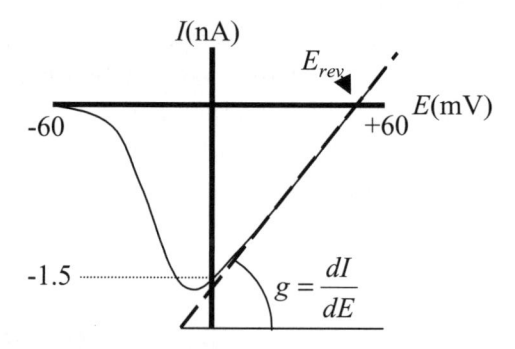

Figure 5.11 Maximum slope conductance is the maximum ratio between current and voltage, as seen from the reversal potential. In this example a 50 mV driving force results in 1.5 nA of inward current. Using Ohm's law, $g_{max} = 30$ nS

Extrapolation to find the reversal potential

If the reversal potential does not lie in the voltage range where the conductance is active, then it can be found using extrapolation of the linear part of the I/V relation or using a tail current protocol (see next Section) if appropriate. Extrapolation would be used if the reversal potential lies outside the voltage range where the cell is 'happy'. This range is -100 to $+60$ mV, or thereabouts, and varies with cell type and duration of the voltage step. Many cells will break down (electrophysiological integrity is lost, resulting in leak and noise) outside this range, so only extrapolation can be used. An example would be a calcium current,

5.2 VOLTAGE-ACTIVATED CURRENTS

for which the reversal potential is often very positive owing to the large concentration gradient across the membrane (Figure 5.12).

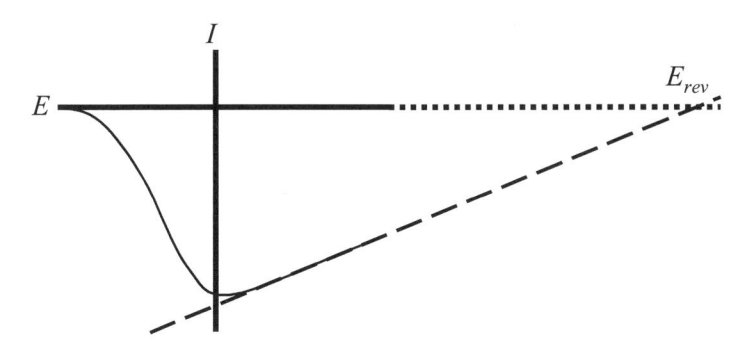

Figure 5.12 Reversal potentials that lie outside the viable membrane potential range can be found using extrapolation from the linear part of the I/V relation

The tail current protocol

A tail current protocol can be used if the reversal potential lies within the viable range but the conductance is not active there, i.e. the channels are normally closed. The trick is to activate the current and then step to a series of potentials near the reversal potential. The active current will shut down, but this takes time and the extinguishing current will leave a tell-tale trail indicating the direction of current flow. This tail current is a function of the initial current, the inactivation characteristics of the channels and the driving force. Tail current protocols are therefore often used to study inactivation (see Section 5.2.7). A typical example of an application would be in the study of a voltage-dependent potassium current such as an outward rectifier. The standard step pulse protocol results and accompanying I/V relation are shown in Figure 5.13.

The reversal potential for this conductance can be found by extrapolation from the linear part of the I/V relation as shown before. However, this would be an inaccurate method because perhaps only two data points seem to be on this line. A tail current protocol allows reversal of current to be seen directly. Firstly, the potential is stepped to a value that evokes a good-size current, e.g. $+20$ mV. The second step is to a series of potentials near the expected reversal potential, in this case between -50 and -100 mV. At these potentials the channels will close quickly but not

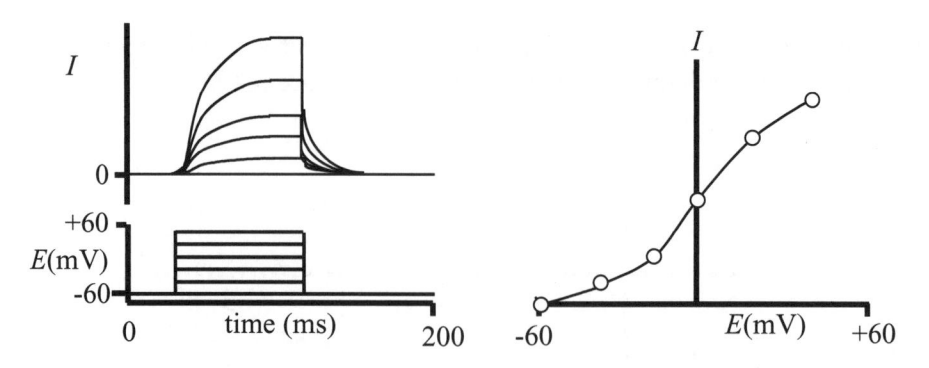

Figure 5.13 The outward current responses to a standard voltage step protocol are shown in the left-hand panel, with the maxima plotted versus step potential in the right-hand panel

instantaneously, so a tail current will be visible (Figure 5.14, left-hand panels). The tail current is a function of initial voltage step, inactivation characteristics and driving force. Thus, assuming that the first two parameters are constant, the tail currents show the reversal potential accurately to within one increment of voltage step (Figure 5.14, right-hand panel).

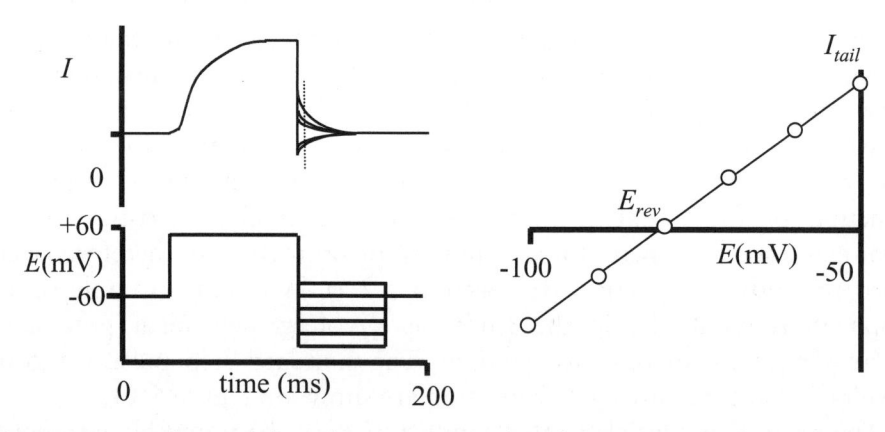

Figure 5.14 Application of a tail current protocol. Tail current amplitudes are measured at the dotted line in the left-hand panel and plotted versus step voltage in the right-hand panel

A very quick way to determine reversal potential is using a voltage ramp. This method is explained in more detail in the Sections on non-voltage-dependent conductances (Section 5.3) and single-channel modes

5.2 VOLTAGE-ACTIVATED CURRENTS

(Section 6.3.2). Voltage ramps are less suitable for voltage-activated conductances because the continuously changing voltage and consequently changing activation and inactivation of the conductance can make the setting of experimental parameters and the interpretation of the results difficult.

5.2.6 Determination of relative permeabilities

Many ion channels show a considerable degree of ion specificity where the ratio of permeabilities (or conductances) favours one ion species over others by orders of magnitude. The reversal potential for the currents through these channels is very near the equilibrium potential for the main permeant ion. In some cases this is sufficient evidence. For example, potassium currents reverse at relatively negative potentials, such that the only equilibrium potential near enough is that for potassium. In most other cases things are not so straightforward and the reversal potential can be due to a mix of permeant ions. Their relative contributions can be estimated in ion substitution experiments. The method of analysis can result in vastly different outcomes so it is important to define relative permeability as conductance *ratios*. One experimental approach ignores the fact that ion species might not behave independently. Ion substitution then involves elimination of ion contributions as best as possible and lists conductances for individual ion species. The conductance then can be described in terms of separate ion conductances. If these do not add up to the total conductance, then the assumption that the ions behave independently is false.

Another approach acknowledges that ion species might influence each other. The conductance g of an ion channel population can be considered to be made up of the conductances for individual ion species

$$g_{total} = g_{K^+} + g_{Na^+} + g_{Cl^-} + \ldots \tag{5.1}$$

Each conductance is built, according to Ohm's law, from the current contributed by the ion species over the driving force for that ion (at the potential the current was recorded). In a given experiment the driving forces are controlled and the total conductance is known, so changes in individual driving forces will be reflected in the total conductance proportional to the relative permeability for a particular ion species. If the changes in ion concentrations are not too large, then the influence of the different ion species on each other's contribution to the total conductance can be considered unchanged. A series of ion substitutions can then determine the relative conductance contribution of each ion species to the total conductance. For a detailed classic example, see Hille (1972).

A large group of channels permeable to several ion species is the group of non-selective cation channels. These include receptor-operated channels such as ionotropic glutamate and nicotinic acetylcholine receptors, and some second-messenger-operated channels. It is often physiologically important to know if these channels are permeable to calcium ions, and ion substitutions can help to establish this (Figure 5.15).

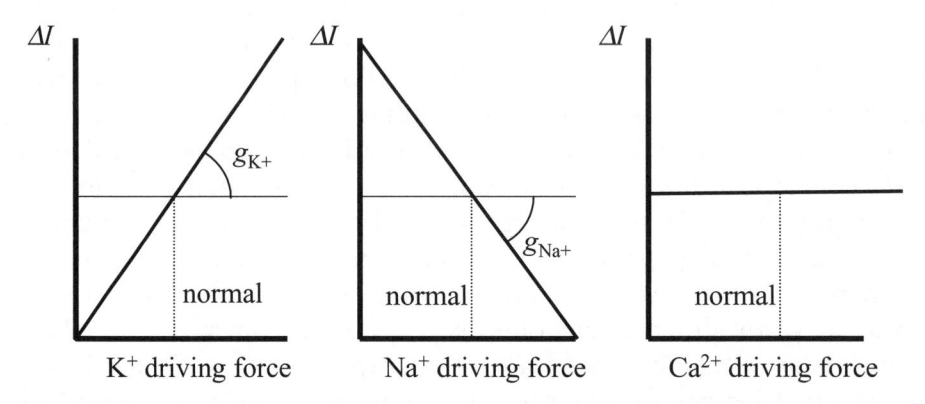

Figure 5.15 An example of a current with a reversal potential of 0 mV that is strongly influenced by changes in the potassium and sodium driving forces but not the calcium driving force

5.2.7 Activation and inactivation studies

Voltage-dependent ion channels activate (open) in response to membrane potential changes; inactivation also can be driven by the membrane potential. In addition, the speed and rate of activation and inactivation are physiologically important characteristics of an ion channel population. Both voltage dependency and kinetic parameters can be studied in pulse protocols.

Activation curve

The activation curve (percentage of maximum conductance versus membrane voltage) is a good way of charting the voltage dependency of

5.2 VOLTAGE-ACTIVATED CURRENTS

activation and can be derived directly from the I/V curve. We saw in Section 5.2.5 that the maximum conductance of a population of ion channels often coincides with the linear part of the I/V relation. If the whole I/V relation is scaled to that particular maximum conductance, then an activation curve can be constructed (Figure 5.16). Note that the activation curve does not indicate the *mechanism* of voltage dependency (see next Section).

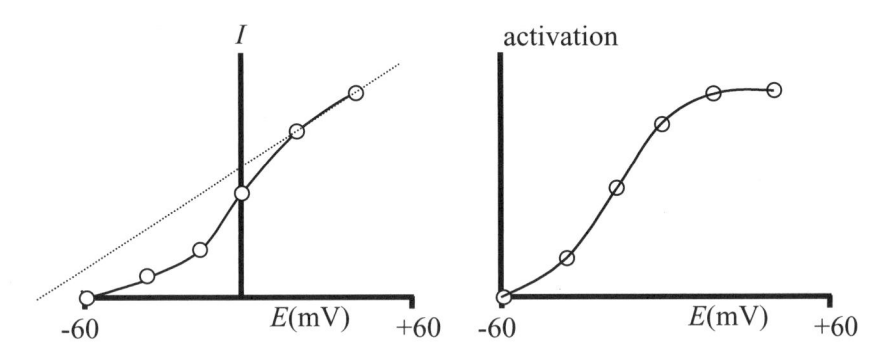

Figure 5.16 Construction of an activation curve (right-hand panel) from the I/V relation of Figure 5.13 (left-hand panel). Each point is the ratio between the observed current and the expected current if the conductance was maximal. The y-axis scale can be conductance or percentage maximum conductance.

Activation and inactivation kinetics

Voltage-dependent ion channels often activate or inactivate upon a membrane potential change because one or more conformational changes occur in the channel protein that open the channel pore. These conformational changes happen very quickly (instantaneous for our purposes) but they occur stochastically, i.e. within a population of channels there is the probability of finding a channel open. The chance depends on the conditions and the channel characteristics. In the case of voltage-dependent activation, the probability of finding a channel open after a voltage step increases immediately after the step and reaches an asymptote. This curve has exactly the same shape as the current (but now with an open probability scale on the y-axis), although the absolute probability scale cannot be read directly from a current trace: it requires single-channel data (see Chapter 6). The shape of current activation and inactivation depends on the characteristics of the channel and is often exponential in nature.

Mathematical fitting of activation and inactivation can either provide precise descriptions of current kinetics that can be compared with data obtained at single-channel level, or reveal multiple-channel populations. Fitting of exponential functions can be done using the larger patch clamp software packages or statistical software.

There are instances where voltage-dependent activation and inactivation occur through other mechanisms. For example, some ion channels, such as inward rectifying potassium channels, are partly or completely blocked at certain membrane potentials by ions. Changing the membrane potential can unblock these channels. This is a very fast process and looks somewhat like an increase in leak (Figure 5.17). As ever in science, it is important to keep an open mind when studying a new preparation and not to over-interpret your data.

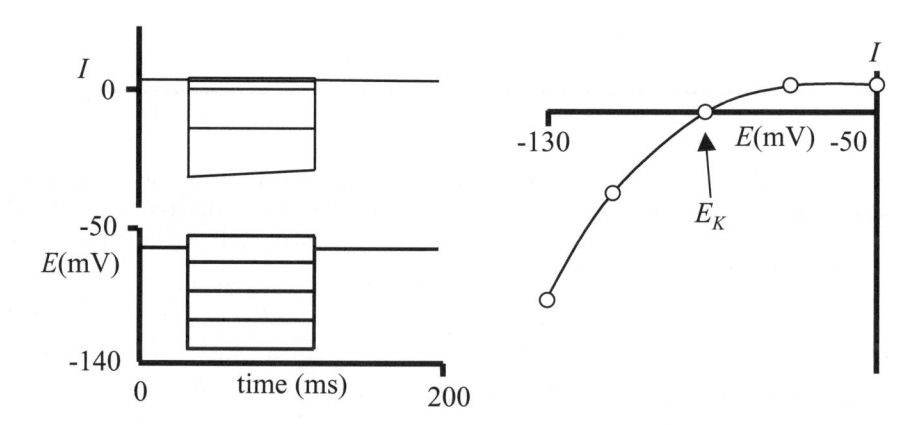

Figure 5.17 Gating of inward rectifying potassium channels is controlled by G proteins, but hyperpolarisation relieves magnesium and polyamine block. In a step protocol this has the appearance of an increase in leak current at very negative membrane potentials

Protocols with a prepulse

The term inactivation is used in this book simply as a decrease in conductance, but in the literature it is also used specifically for when inactivation lies in the same voltage range as activation. The outward current activated in Figure 5.13 will stay activated until the membrane potential is changed again. The voltage range for inactivation of this current seems therefore completely separate from the activation range. In contrast, the inward current from Figure 5.10 quickly shuts down after

5.2 VOLTAGE-ACTIVATED CURRENTS

activation, even though the potential remains the same. So when the channels open, a mechanism is put into action to close the channel again. The special closed state in this situation is sometimes referred to as the *inactivated state*, and is associated with a separate gate from that involved in the activation process. For example, the current from Figure 5.10 could be mediated by TTX-sensitive sodium channels, which indeed possess two different gates (Figure 5.18). Closing of the inactivation gate is time dependent, whereas opening it is voltage dependent: relief of time-dependent inactivation can only take place by hyperpolarising the membrane.

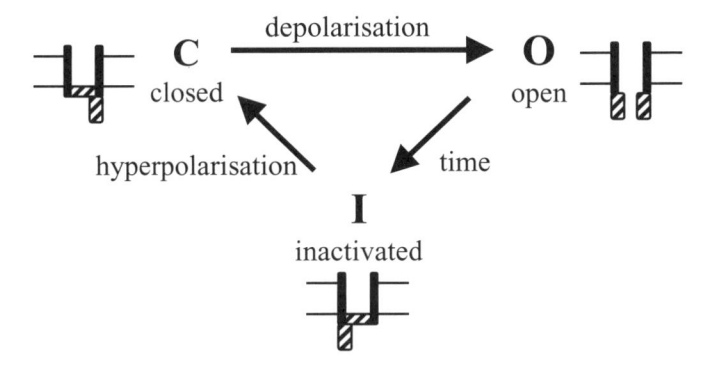

Figure 5.18 A sodium channel has separate activation and inactivation gates operated by membrane voltage and time

The closure kinetics can be observed directly in the whole-cell currents, although it is not straightforward to discriminate between (slow) inactivation gate closure governed by voltage or by time spent in the open state. The voltage-dependent opening of the inactivation gate cannot be observed directly because the channel's activation gate is mostly closed at potentials where inactivation is removed. There is, however, a method to quantify the voltage dependency of inactivation opening: the inactivation curve. The chance of finding a TTX-sensitive sodium channel with an open inactivation gate increases with hyperpolarisation. If a constant depolarising pulse (opening the activation gates of the sodium channels) is preceded by a variable prepulse, then the evoked sodium currents will differ in amplitude, depending on how many of the channels were locked in the inactivated state during the prepulse. If the prepulse is sufficiently long, then the proportion of channels in the inactivated state can be considered time independent and is therefore termed *steady-state inactivation*, denoted by the parameter h_∞. This parameter is derived from the inactivation parameter h from Hodgkin and Huxley's squid giant axon sodium current kinetics. An example of a protocol with prepulse is given in Figure 5.19. The use of the prepulse is not restricted to detailed study of inactivation. It is generally applied, as a pulse of constant amplitude, to maximise the amplitude of currents mediated by inactivating channels such as TTX-sensitive sodium channels and several types of voltage-dependent calcium channels.

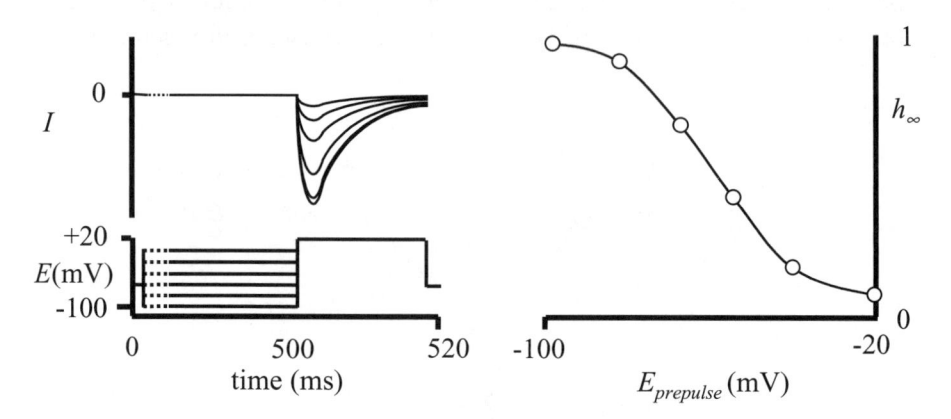

Figure 5.19 The steady-state inactivation parameter h_∞ can be derived from a step protocol with prepulses. The evoked currents are scaled to the maximum current and plotted against steady-state potential. The parameter indicates directly the fraction of channels that can be recruited for activation

If there is an overlap between the activation curve and the inactivation curve, then there will be a small, constant current in the area of overlap (Figure 5.20). This is called a window current and can be physiologically significant.

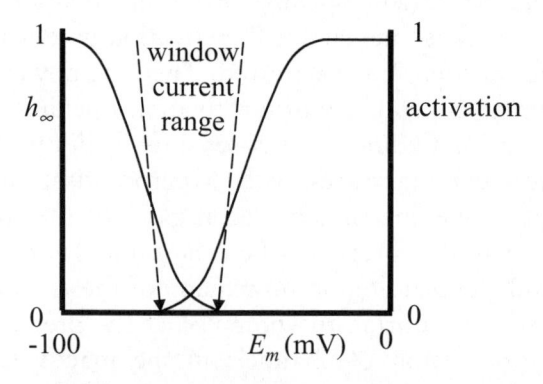

Figure 5.20 Overlap between activation and inactivation curves creates a range of membrane voltage where a small, continuous current is present

5.3 Non-voltage-activated Currents

5.3.1 Introduction to continuous recording

The distinction between voltage-activated and non-voltage-activated conductances is not always clear-cut. Most classic voltage-activated channels can be modulated by intracellular factors, which can strongly influence their functioning. Even if one considers only the primary gating factor, it is not clear where to classify, for example, the large-conductance calcium-dependent potassium channels, because they require both depolarisation and a rise in intracellular calcium to be activated. For the practical purpose of this Chapter, non-voltage-activated currents are simply those that are studied using continuous recording instead of step protocols, although they can still be influenced by voltage. In any case, holding potential plays a role in determining the driving force for the ions passing through the channels.

Continuous recording is straightforward using the main software packages, but there can be some awkward dilemmas to solve. Most non-voltage-operated channels, such as second-messenger-operated channels (SMOCs), activate at a much slower time scale than those activated by voltage steps, so the sampling rate and filter frequencies must be chosen accordingly. This is particularly difficult in single-channel recording, because channel openings are relatively brief and receptor activation and signal transduction can be orders of magnitude slower. It is also harder to set the amplification factor because it is often not possible to use trial and error to find the ideal value, as is usually the case with pulse protocols. For these reasons many workers use DAT recorders, or similar, to provide a continuous, raw record of the experiments that can be analysed off-line. Another advantage in using this type of recording is that the computer-driven data acquisition and voltage clamp is freed up for other purposes, such as reversal potential experiments (see Section 5.3.2).

A good routine of labelling must be established in continuous recording. The start of an activation can involve turning a tap or other means that are not necessarily reflected in the recording. If you use a chart recorder and have a pen, then there is no problem with labelling. Some software and AD converter combinations allow entry of a digital time marker, perhaps with a written comment, to be stored with the experimental trace. A DAT recorder has two channels, so the second channel can be used for digital or even verbal labels. Whatever is used, labelling should be integrated thoroughly in the experimental routine.

5.3.2 Determination of reversal potential using voltage ramps

The methods explained in Section 5.2.5 to determine the reversal potential of the conductances under study are often not practical for non-voltage-activated currents. To acquire a reasonable accuracy, a stimulation would have to be repeated many times at different holding potentials, possibly on different cells. Cells would need to be clamped at unhealthy potentials for extended periods. Instead, the relative slowness of many non-voltage-activated conductances can be exploited by using quick voltage ramps. The idea is that a voltage ramp is applied to a cell before, during and after activation of a non-voltage-activated conductance. Similar to the principle of P/N subtraction (Section 5.2.2), the averaged trace before and after the response is subtracted from the trace recorded during the response. The result should reflect the I/V relation of the non-voltage-activated conductance (Figure 5.21). In addition to the reversal potential, the relation will

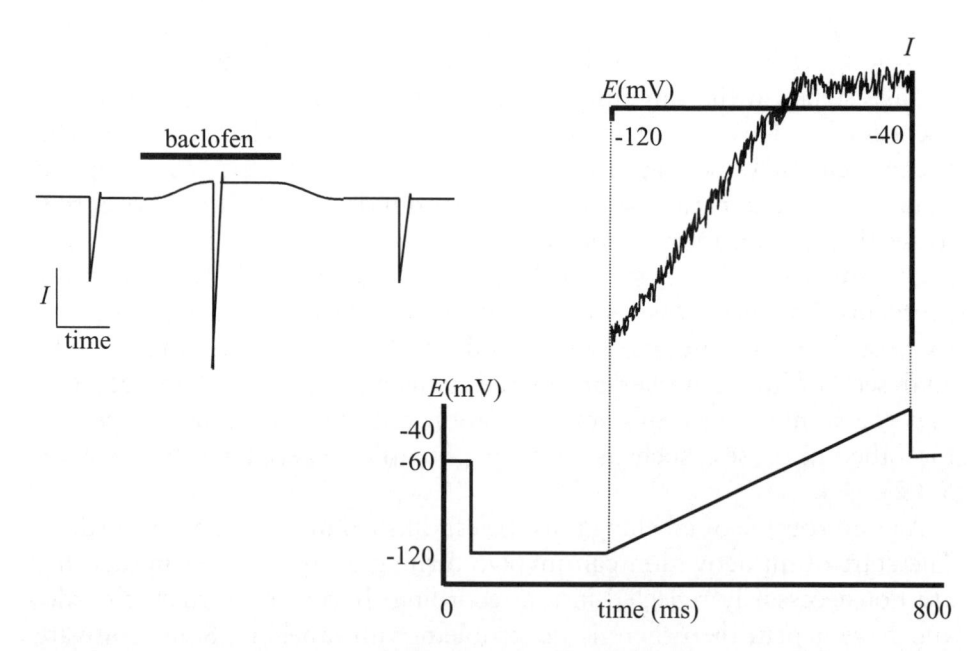

Figure 5.21 A current response induced by the GABA$_B$ receptor agonist baclofen is probed by a voltage ramp (left-hand panel). The ramp currents obtained before and after the $GABA_B$ response are averaged and subtracted from the current obtained during the response. The result in the right-hand panel shows the profile of an inward rectifying potassium conductance (of Figure 5.17). The current scale can be of the order of hundreds of pico-amperes

5.3 NON-VOLTAGE-ACTIVATED CURRENTS

also show any voltage dependency of the conductance. If the result is a straight line, the conductance is not voltage dependent, but if there is a curve, then the current rectifies; this is an important characteristic.

There are several issues that must be considered when applying ramp protocols:

- The response must be sufficiently slow compared with the voltage ramp, so that the response can be considered constant in magnitude while the ramp takes place.

- The ramp must be slow compared with the RC response of the circuit, the filter settings and the sampling rate. All of these can introduce delays or other distortions which can make it invalid to match the current response to the applied voltage ramp. As an indication, a voltage ramp should be 500–1000 ms in length when the RC of the circuit is about 1 ms (e.g. $C_m = 50$ pF and R_{series} is 20 MΩ), the low-pass filter cut-off is 2 kHz and the sampling rate is 4 kHz (see also Section 2.3.6). A check for the occurrence of artefacts caused by delays can be made by recording two ramps: one from negative to positive potentials, and a reversed one. They can be applied in succession or separately. The corrected current responses ideally should be mirror images.

- These experiments span a relatively long time. A small change in capacitive transient can create a big artefact in the recording, so it is prudent to use a settling time after the initial step before the ramp starts (as in Figure 5.21), to bypass most of the capacitive transient.

- Cells suffer from being clamped at extreme potentials, and non-voltage-activated currents can be drowned out by voltage-activated currents. It might be necessary to use channel blockers and/or change ion concentrations for best results.

Voltage ramps can also be used in single-channel recording, with the added advantage that they can be very fast due to the very small time constant of a patch of membrane. Single-channel recording is the subject of the next chapter.

6
Single-channel Protocols and Data Analysis

6.1 General Single-channel Practice and Analysis

6.1.1 Practical notes

It was explained in Chapters 2 and 4 that single-channel recording can be performed in three different modes: cell-attached patch, outside-out excised and inside-out excised. Each of these has its own advantages in terms of the possibility of experimental manipulation. All single-channel modes, however, result in single-channel data that are very characteristic and require special types of analysis. Single-channel recording involves the recording of currents through conductors (channels) with conductances of the order of 2–500 pS, which is equivalent to resistances of 2–500 GΩ. It is clear that the seal resistance, which is parallel to the channel, must also be very high in order to 'see' currents contributed by the channel conductance. In general, the seal resistance for single-channel recording must be of considerably higher quality (size and stability) than for whole-cell recording.

Once the desired configuration is established, amplification and filtering of the signal should be optimised so that changes of the order of a few pico-amperes can be seen. Although there is no whole-cell capacity, normally inconspicuous capacitances such as pipette capacitance can cause substantial transients in response to voltage steps due to the high magnification factor and also because recording sometimes must be performed on fast time scales. At this stage, therefore, some time must be spent in minimising these transients on the output signal. The same test pulse used for establishing the seal can be used for this. It is then time to establish how many channels have been captured and what type they are. This is largely a matter of luck. For most studies it would be ideal if there is only one channel in the patch, because it greatly simplifies data analysis if all observed activity can be attributed to one unit. The second most preferable situation is if there are more channels in the patch but they are all of the same type. If several different channels are present, analysis quickly becomes very complicated or impossible. The patch initially will be clamped at a near-physiological resting membrane potential because this is usually the most stable voltage. Note that the pipette potential must be set at very different values for the same patch potential of −60 mV, for example, depending on the configuration: 0 mV for cell-attached patch, −60 mV for outside-out patch, and + 60 mV for inside-out patch (more about conversion from pipette to patch potential will be given in Section 6.2) (see Figure 6.1). Single-channel activity could be visible without any manipulation at all, and would show as stepwise current deflections (transitions) as in Figure 4.9.

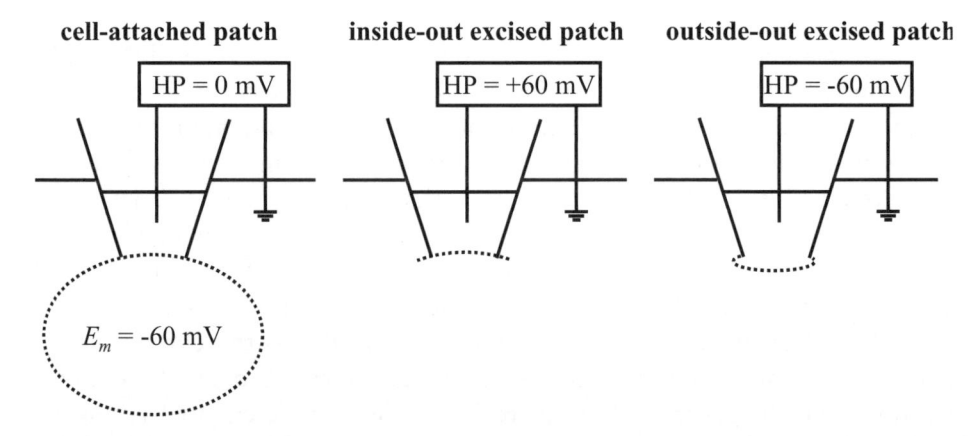

Figure 6.1 The configuration must be taken into account when setting the holding potential (HP). To expose the patch to a potential of −60 mV, a different HP must be applied in each configuration

More often than not there is no spontaneous channel activity visible, so many workers perform a quick voltage trawl on a new patch by applying a number of robustly depolarising steps of constant length and amplitude, or by manually exposing the patch to a voltage range as wide as the patch will take without breaking up. The latter is a very rough manipulation that is perhaps best saved for end-of-experiment playtime. If there are single channels that can be persuaded to open by this treatment, the openings and closures will appear as the characteristic single-channel current transitions. Two important points here are:

1. There can still be voltage-dependent channels in the patch if the initial voltage trawl does not reveal anything. Among the many possible reasons for this are: it could be that the applied step was wrong in amplitude for opening or for sufficient driving force, or too long or short in length. It could be that manual changes are too slow if the channel inactivates very quickly, etc. Rejection of a patch at this stage is only justified when the experimenter has enough experience with the preparation to know that his or her manipulations should definitely cause transitions of the channel under study but none are visible, or if many (different) channels appear and selective activation is impossible, so that the data become impossible to analyse.

2. There is no direct evidence at this point to indicate that there are no *open* channels in the patch. Circumstantial evidence from higher noise levels, different resting current and lower apparent seal resistance than average are hints. By the same token, a transition to the closed state of a channel is accompanied by a decrease in noise, but this might not be easily visible (more about this in Sections 6.1.2). Note that the direction of current (inward or outward) does not *per se* indicate which current level represents the closed state.

Once a patch is isolated, the channel of interest can be spontaneously active or must be stimulated by voltage changes, receptor activation or intracellular factors. In the case of voltage changes, voltage step or ramp protocols are applied (see Section 6.3), whereas in the other cases recording is continuous. This will be discussed in Section 6.2. There are many commonalities in analysis of the data, which are explained below, and specific aspects of the two data types will be discussed in their own sections.

6.1.2 Amplitude analysis

Analysis of single-channel data concerns in the first instance a description of the channel in terms of open/closed times (x-axis) and current amplitude (y-axis). Current amplitude can provide information on single-channel conductance and ion specificity of the channel. There are two common ways of measuring single-channel amplitude:

1. *'Manually'*. The difference between two current levels in raw data is measured. In practice, the middle of a noise band should be taken as datum for each current level (Figure 6.2). The advantage of manual measurements (usually on-screen by cursors) is that they can be taken from clearly identified and associated current levels within a region of the data. The automated method leaves the experimenter less in control, as explained below.

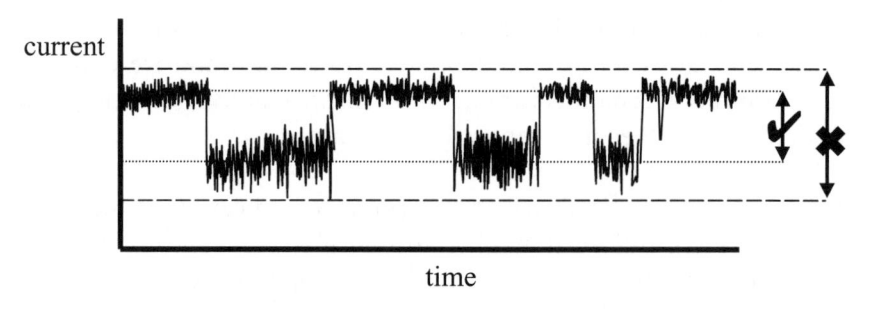

Figure 6.2 Channel amplitude should be measured between the centre line of noise levels, not as the maximal difference

2. *Automatically.* An amplitude histogram can show the levels of current present in the data. A discussion of amplitude histograms follows.

Amplitude histograms

Digitised data are stored as an array of amplitudes – one amplitude value for each sample. (This implies that each amplitude value represents a unit of recording time, and counting the number of samples divided by the sampling rate will render the duration of the recording. This property is used in dwell-time analysis; see Section 6.1.4.) The amplitudes are sorted by size in a number of bins (30–200 is reasonable) that cover the amplitude range of the data (Figure 6.3). Current levels can be identified by peaks in the histograms. The curves that form the peaks are the normal (Gaussian) distributions of the noise at the current levels. All permutations of channels, open or closed, will each result in a current level and a

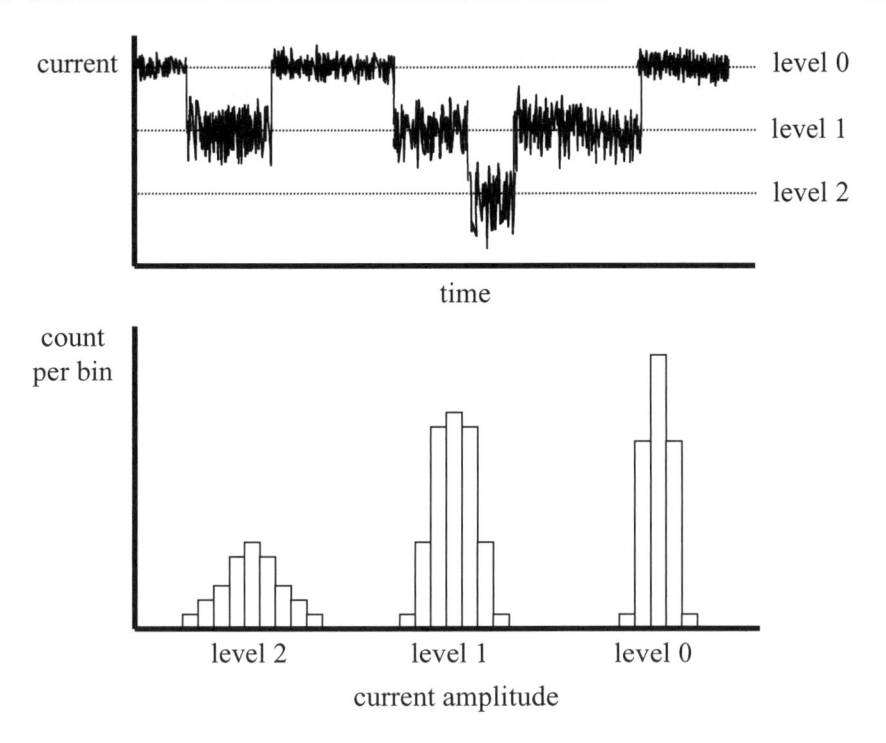

6.1 GENERAL SINGLE-CHANNEL PRACTICE AND ANALYSIS 145

Figure 6.3 Amplitude histograms show current levels as Gaussian (normal) distributions

subsequent histogram peak. If the channel activity is caused by one or more channels of the same conductance, then the channel current amplitudes can simply be read as the distance between any adjacent peaks (Figure 6.3), but if the recording contains the activity of more types and numbers of channels, then the situation quickly becomes very complicated because the current levels of different permutations are likely to get very close to each other. For example, a patch containing two small and two large channels has tens of permutations for stacking opening events, of which several may coincide! The data would be impossible to analyse. This is partly why patches containing several channels of different types are often no good.

Notes on amplitude histograms:

1. To measure the distance between two peaks they obviously need to be of the same order of magnitude. If a channel rarely opens, then the 'open' peak might be nigh on invisible in the histogram. This can be solved by selecting for the histogram a part of the data that contains

the rare openings and excluding the rest, so that the 'open' and 'closed' peak are similar in magnitude. It is perhaps superfluous to say that these selected histograms should not be used for dwell-time analysis!

2. Amplitude histograms work best with lots of data and a good signal-to-noise ratio, but the noise can be a help too. It was mentioned in Section 6.1 that one way of telling what the 'open' and 'closed' current levels are is to measure the noise. The 'open' level will be noisier due to increased thermal noise through the channel. The amplitude histogram provides a very good picture of noise in the form of peak width, hence the wider histogram will likely represent current levels with open channels (Figure 6.3).

3. An amplitude histogram with a set of perfect Gaussian distributions would be obtained if the transitions were infinitely fast. However, the low-pass filter that is always needed in these recordings slows down the transitions in the recorded data (not in reality), so that the transitions will occasionally be visible in the data and show up in the histogram as a bridge between peaks (Figure 6.4). The height of this bridge is a function of the number of transitions and the filter cut-off frequency. The filtering will also depress the amplitudes of very short openings or closures. It is clear from these considerations that small channel currents (peaks close together) and many fast openings require high cut-off frequencies and fast sampling speeds. This is even more apparent in dwell-time analysis (Section 6.1.4).

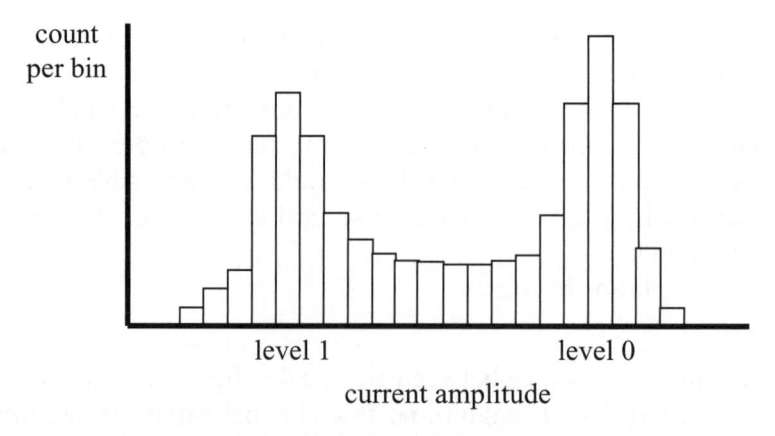

Figure 6.4 Filtering artefacts create a 'bridge' between current levels as transitions are slowed down in the recording

6.1 GENERAL SINGLE-CHANNEL PRACTICE AND ANALYSIS

4. The peaks in amplitude histograms can be fitted using iterative computer algorithms, which, if successful, provide precise and occasionally surprising summative data. Fitting routines are included in the big patch clamp software packages and in some statistical analysis packages. They are definitely not to be relied on blindly, however, and results should always be checked graphically. The fitting routines can have variable levels of automation where the experimenter can choose to enter seed values and fix some of them. This implies some judgement skills and practice.

5. Some ion channels have 'sub-conductance states' where it is thought that a single-channel protein can have more than one open conformation, with different conductances. These can appear in amplitude histograms but it is often difficult to tell if there is a sub-conductance present or a second, smaller channel. The presence of two signs in the data would point towards a sub-conductance state (Figure 6.5):

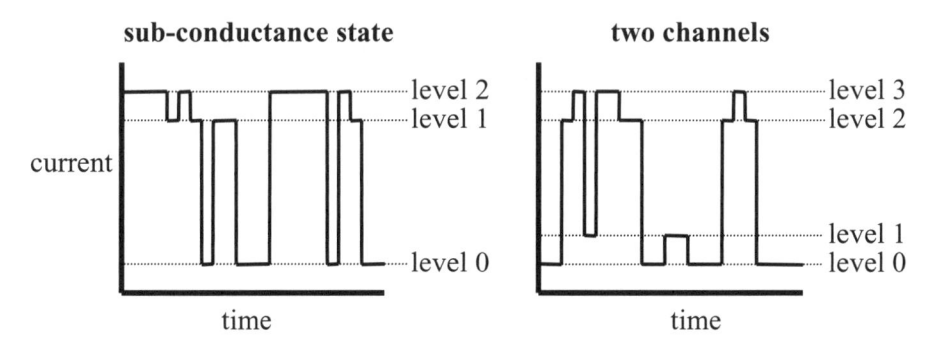

Figure 6.5 Sub-conductance states can be distinguished from a small channel by the absence of a current level and transitions between both the sub-conductance state and baseline and the full conductance level and baseline. In the left-hand panel level 1 represents the sub-conductance state, whereas in the right-hand panel level 1 represents the open small channel, level 2 represents the open large channel and level 3 is when both channels are open

- There are transitions from baseline to or from both the large current amplitude and the smaller one. In the case of two channels, it is much more unlikely that they will both open or close simultaneously (within the resolution of the recording).

- If the sub-conductance state is not exactly half the full conductance, then the amplitude histogram will show three unequally spaced peaks. If there were two different channels, there would be four peaks.

Single-channel current-voltage relationships

Current–voltage (I/V) relations can be constructed if channel current amplitudes can be obtained at different patch potentials. The relation provides the most basic information about the channel under study:

- *Single-channel conductance and rectification.* Given that conductance is the reciprocal of resistance, Ohm's law says that the slope of the relation will directly provide this vital characteristic. If there is rectification (the relationship is not a straight line), then the maximum slope should be taken and a description of the curve should be added to the characterisation, perhaps as a range of chord conductances. However, rectification at the single-channel amplitude level is rare.

- *Reversal potential and permeant ions.* Analogous to whole-cell I/V relations, the reversal potential of the single-channel current helps to determine which ion species pass through the channel. For example, if the reversal (patch) potential is -80 mV in standard solutions, then the channel is very likely to be a potassium channel, because potassium is the only ion that has an equilibrium potential near that value under those conditions (see Figure 6.6). As mentioned previously, care should be taken here that, whatever the configuration, the potential *across the patch* should be considered and not pipette potential or anything else.

If a look at the noise levels is not decisive in determining which is the direction of the current and what the different levels represent, then the I/V relation can help. Because the steepest line represents the largest conductance, this line represents the open level. The resting conductance (or seal resistance) can also be determined using the I/V relation.

6.1.3 Event detection

Time analysis of single-channel data depends largely on successfully identifying transitions and cataloguing them. This involves scanning the data for transitions or 'events'. With very stable data, not many transitions per unit of time and a good signal-to-noise ratio, event detection is easy

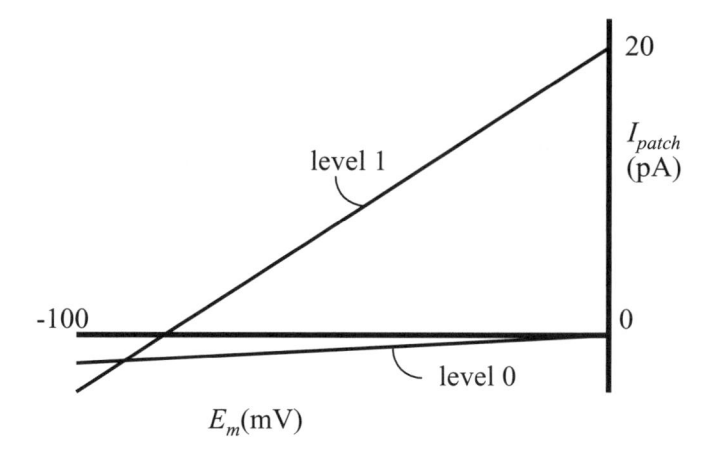

6.1 GENERAL SINGLE-CHANNEL PRACTICE AND ANALYSIS

Figure 6.6 Example of an I/V relation from a patch containing a potassium channel. The baseline (level 0) reverses at 0 mV and has a resistance (1/slope) of 50 GΩ. Level 1 reverses *with reference to baseline* at −90 mV and has a conductance (= slope) of 244 pS, indicating a large channel

and can be automated with some confidence using simple thresholds. Data between transitions are usually stored as pairs of numbers, indicating average current or level number and the time spent there (dwell time). When plotted like this, the result is an idealised trace (Figure 6.7). Most patch clamp software incorporates event detection using thresholds.

Event detection becomes more difficult with noisier data, a high density of transitions (particularly successive transitions very close together in time, i.e. very short openings or closures) and a moving baseline (drift). With noisy data, a simple threshold cannot be applied because there is no 'safe' amplitude region that lies between the two current levels. Before resorting to detection by eye, there are several strategies that can be followed to rescue the automation. Two possibilities are:

1. *Off-line filtering.* If the number of fast successive transitions is low (see further in this section) and the noise from the different levels overlaps only slightly, then stronger low-pass filtering (lower cut-off frequency or steeper cut-off) might create a window for a threshold.

2. *Two-threshold detection.* Different thresholds can be assigned to different transitions, i.e. the threshold for a transition from level 0 to level 1 can be different from the threshold assigned to the transition from level 1 to level 0. In this way, each threshold can lie safely outside the

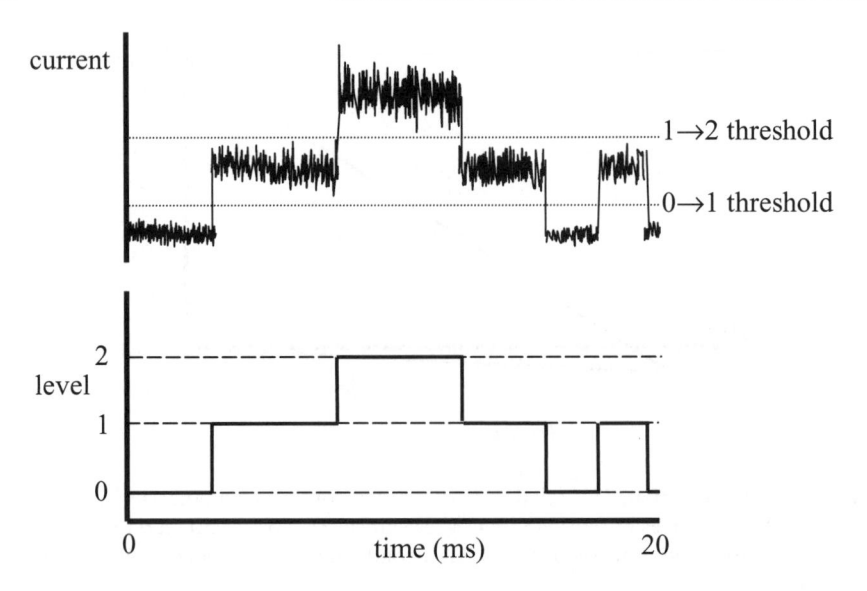

Figure 6.7 Event detection using single thresholds between levels. The result is an 'idealised trace' (lower panel) that can be stored as a series of level number/dwell time pairs. This trace could be stored as 0/3.0, 1/4.5, 2/4.6, 1/3.1, 0/1.9, 1/1.8 and 0/1.1, which is a considerable data reduction from the original trace of 1000–2000 points

level from which it is supposed to detect a (departure) transition (Figure 6.8).

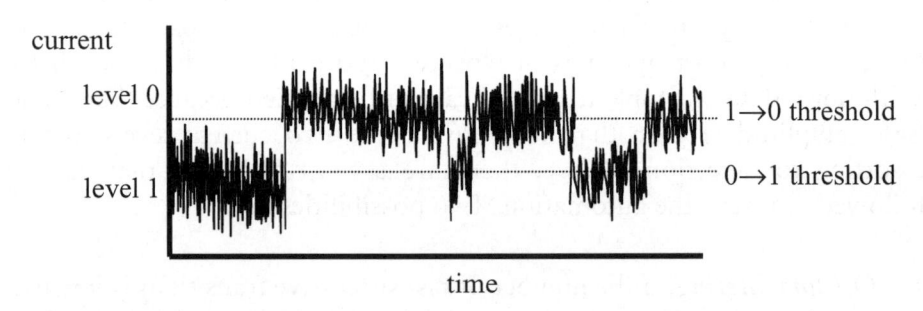

Figure 6.8 Event detection of noisy data can still be automated using multiple thresholds. The setting criterion for each threshold is that it lies outside the departure level domain and well in the arrival level domain

Very short openings and closures result in incomplete-looking or even invisible transitions in the data due to filtering, either by the set low-pass filtering or any uncompensated capacitance in the circuit (see Section

2.2.3). An incomplete-looking transition can be seen at the end of the trace in Figure 6.2. This artefact is inevitable. The consequences are that fast dwell times are missed and some slower dwell times are misinterpreted as being longer than they really are. The former is easily dealt with by ignoring dwell times faster than, for example, half the cut-off frequency of the low-pass filter used (<1 ms dwell times for data filtered at 2 kHz). The error in slower dwell times, however, is not so easy to overcome. It follows that it is worth checking the data carefully for incomplete transitions and performing some detection by eye if necessary, even in otherwise 'clean' data. This problem is another argument for recording data with minimal filtering: what is filtered out before data storage can never be retrieved, and filtering can always be done off-line.

A shifting baseline creates havoc with automated event detection using set thresholds. If the drift is very slow, it can be overlooked by the

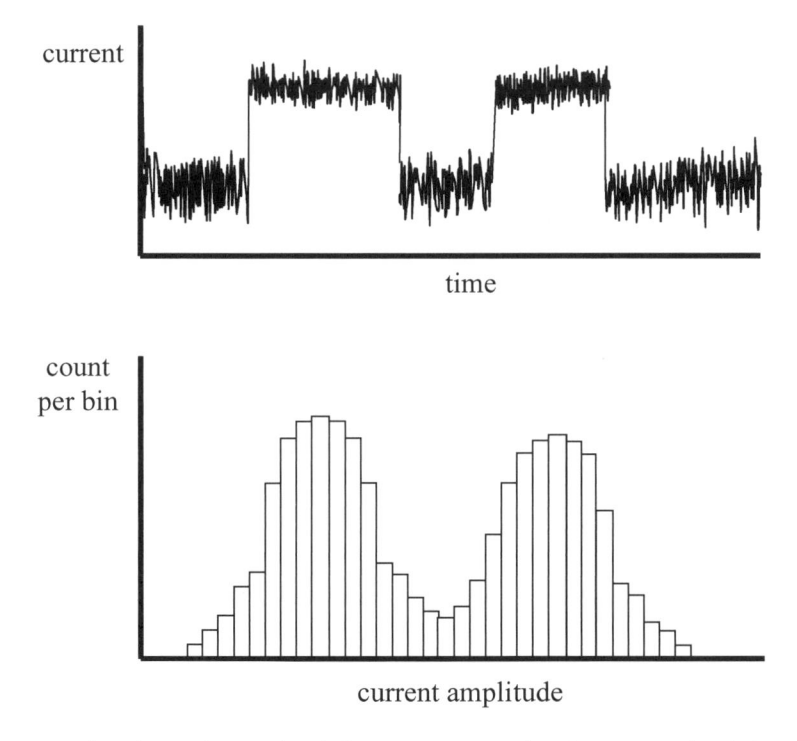

Figure 6.9 If peaks in the amplitude histogram are wider than a sample of the raw data suggests, then slow drift in the recording could be a factor. A single threshold event detection without baseline tracking will not work here

experimenter. One way of looking for drift in the analysis stage is to study the amplitude histogram. If the noise of the current histograms is clearly larger (wider histograms) than is apparent from the raw data, then drift is a likely cause (Figure 6.9). Event detection in some patch clamp software allows for drift correction. During the event detection process, consecutive average current values of the current levels are compared and, if they are different, thresholds are adjusted accordingly. In this way the appropriate local thresholds are applied throughout the data. The main caution with this method is that sudden changes in the data can throw the process off the rails, so inspection by eye is essential. In light of the above, this rule can be generalised for nearly all single-channel event detection.

6.1.4 Dwell time analysis

Analysis of dwell times (sometimes called sojourns) is the study of transition patterns. Mathematical modelling of dwell times can help to describe accurately the behaviour of a channel, and also point to physiological mechanisms. The simplest and perhaps most important parameter in dwell time analysis is the open probability, i.e. the chance that the channel will be found open in a given set of conditions. It is the open probability, $P(open)$, together with the driving force of the permeant ions, that determines what the current through a population of channels is, and the beauty of single-channel recording is that it allows the two factors to be studied separately.

Open probabilities

Open probability $P(open)$ at a given potential can be obtained from the amplitude histogram if the current amplitudes can be separated adequately, either manually or by fitting with Gaussian functions. Each sample of data is included as an occurrence in the amplitude histogram, so the sum of all occurrences (area under the curve) represents the total recording time. In recordings with only one channel, $P(open)$ is then the area under the curve for the 'open' peak divided by the total area

6.1 GENERAL SINGLE-CHANNEL PRACTICE AND ANALYSIS

$$P(open) = \frac{\sum t_{open}}{\sum t_{open} + \sum t_{closed}}$$

$$= \frac{\sum occurr., open}{\sum occurr., open + \sum occurr., closed} \quad (6.1)$$

or, in integral notation

$$P(open) = \frac{\int f_{open}(ampl)}{\int f_{open}(ampl) + \int f_{closed}(ampl)} \quad (6.2)$$

where $f_{open}(ampl)$ and $f_{closed}(ampl)$ are the Gaussian distribution functions describing the peaks. The situation gets a little more complicated with multiple channels (of the same type) in the patch, resulting in multiple peaks in the amplitude histogram. The counts in the various peaks must then be multiplied by the appropriate factor in view of what they represent. This is illustrated in the following example. In a histogram with four equidistant peaks (suggesting three channels), the baseline peak (level 0) represents all three channels closed. The second peak (level 1) represents two channels closed and one channel open, whereas the third peak represents two channels open and one closed. Under the fourth peak (level 3) all channels are open. The summated closed time of all three channels in the histogram can be calculated as

$$\sum t_{closed} = 3\int f_{level0}(ampl) + 2\int f_{level1}(ampl) + \int f_{level2}(ampl) \quad (6.3)$$

Conversely, the total open time is

$$\sum t_{open} = \int f_{level1}(ampl) + 2\int f_{level2}(ampl) + 3\int f_{level3}(ampl) \quad (6.4)$$

and $P(open)$ can then be calculated as explained earlier. Another way to find $P(open)$ is by using the idealised traces. The dwell times at each current level can be summated and processed in the same way as amplitude histogram peaks. The disadvantage of using idealised traces is that a process of event detection must take place for this.

Two cautions must be stated regarding open probabilities:

1. The number of current levels is not necessarily one more than the number of channels in the patch; there could always be more channels. This implies an inherent possibility of overestimation of P(open).

2. Open probabilities are only meaningful under steady-state conditions. If measured under changing conditions, then P(open) or any other kinetic property should be determined for time intervals short enough to consider them as pseudo-steady-state. This is pertinent to evoked activity such as the rising phase of a voltage-dependent current or an inactivating current. Another method of calculating P(open) for such conditions is presented in Section 6.3.1.

A graph of P(open) versus membrane potential provides a good picture of the voltage dependency of gating, and has the same shape as the whole-cell activation curve (Section 5.2.7) provided that the channel *conductance* is not voltage dependent.

Dwell time distributions and channel modelling

Many short openings in a data set can result in the same P(open) as that for a few but long openings. Hence, the length of dwell times in a data set provides more detailed information about the stochastic gating behaviour of the channel. The dwell times as found in idealised traces can be arranged in dwell time histograms for each current level found in the data set. Kinetic models describing the channel's behaviour can be derived from these histograms. The following is meant to be an introduction to this type of analysis and will consider a patch with only two current levels (one channel) for simplicity.

Changes from the open to the closed state of the channel, and vice versa, represent conformational changes of the channel protein. These changes occur spontaneously (due to thermal movement) but require the protein to hop an energy barrier between the two (or sometimes more) conformations. The time scale of these events is orders of magnitude shorter than single-channel recording and translates at the experimental level to stochastic channel gating. There is a *probability* that the channel will change conformation at any moment in time and that this probability is independent of any gating that went before. This is sometimes referred to as Markov behaviour. As a consequence, the likelihood of a channel remaining in one conformation such as open or closed becomes less with increasing dwell time and hence a distribution that decays with time is expected.

The exact shape of dwell time distributions (Figure 6.10) can be explained by the underlying dynamic equilibrium between the different possible protein conformations, some of which impart an open gate and others a closed gate (ignoring sub-conductance states for the moment; see Section 6.1.2). The simplest situation has the protein switching between the two states (Figure 6.11).

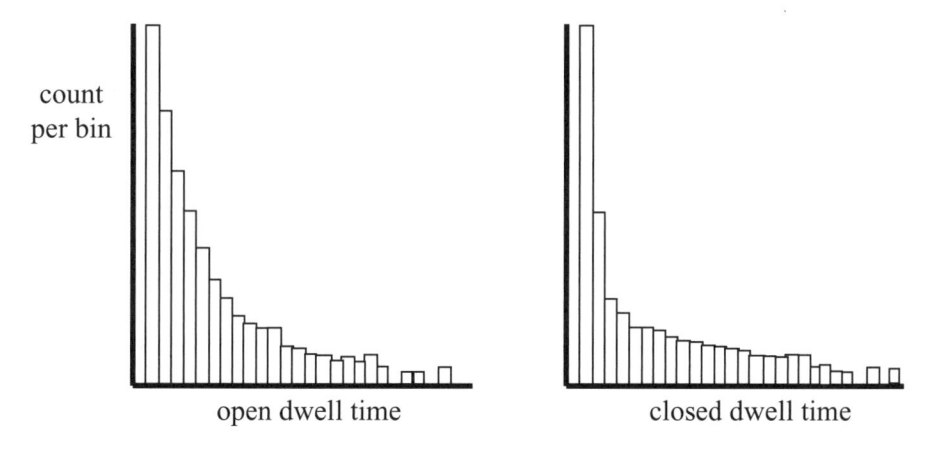

Figure 6.10 The shape of dwell time distributions is exponential. In this example the dwell time distribution for open dwell times is a single exponential, whereas the closed time distribution is the sum of at least two exponentials. Note that the distributions do not start at zero time, because the filter properties of the recording configuration set a lower limit to the dwell times than can be observed reliably (see also Section 6.1.3)

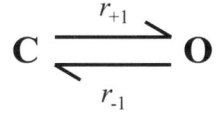

$$C \underset{r_{-1}}{\overset{r_{+1}}{\rightleftarrows}} O$$

Figure 6.11 Simple gating model with one closed and one open state

The situation in Figure 6.11 looks like a chemical reaction, although an important fundamental difference is that, because we only consider one channel, the rates between states are more like transition probabilities than reaction rates. However, the principle is very similar. The probability of transition is determined by the state of the channel and the rate constant(s) describing the transition *away* from that state. On the basis of this model, it can be shown that the resulting distributions of dwell times follow an exponential decay (see Colquhoun and Hawkes, 1995, for detailed explanation)

$$f_{open}(t) = r_{-1}e^{-r_{-1}t} \qquad (6.5)$$

and

$$f_{closed}(t) = r_{+1}e^{-r_{+1}t} \qquad (6.6)$$

Hence, if the exponential functions describing the dwell time distributions can be found, the rate constants can be derived. This in itself provides a good characterisation of the channel's gating behaviour, but only provided that the data are recorded under conditions where the rate constants are indeed constant. The rate constants found under different conditions (e.g. different membrane potentials, presence of a blocker, etc.) can be compared and statements can be made about the dependency of individual rate constants on a certain variable.

If there are more than one open and/or closed states, can they be detected in this analysis? It all depends on the state model. In general, the dwell time distribution for a given state is (single) exponential in nature, as long as transitions away from the state involve other states that are not independent. It follows that only dwell time distributions that involve transitions to independent states will have the shape of the *sum* of different exponentials. A classic example of such a dwell time distribution is the closed time distribution involving a blocker (see Figure 6.12). If the

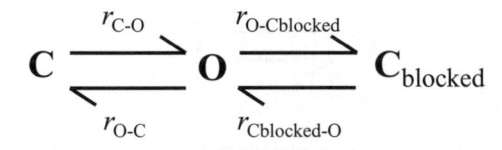

$$C \underset{r_{O-C}}{\overset{r_{C-O}}{\rightleftarrows}} O \underset{r_{Cblocked-O}}{\overset{r_{O-Cblocked}}{\rightleftarrows}} C_{blocked}$$

Figure 6.12 The presence of a reversible blocker introduces a second type of closed state. The open dwell times are terminated from one open state, which results in a single exponential open dwell time distribution. The closed states, however, are terminated by two independent mechanisms, so that the closed dwell time distribution is the sum of two independent exponential functions. The corresponding distributions could be those shown in Figure 6.10

exponentials are different enough, then this can be detected using exponential fitting where the number of exponential functions is not fixed *a priori*. You would then determine the number of exponentials involved in the distribution using maximum likelihood algorithms, which balance improved goodness of fit with additional degrees of freedom for each added exponential. These analyses are part of the larger patch clamp software packages. In addition to identifying sums of exponentials, information can be derived from changes in single exponentials (which might be a product of multiple transitions): it could be that an experimental manipulation points to a 'hidden' state by modifying the dwell time distributions of a state that can be observed.

This concludes the introduction into general single-channel studies. Specific considerations regarding continuous recording and voltage protocol recording are discussed in Sections 6.2 and 6.3.

6.2 Continuous Recording of Single Channels

6.2.1 Data acquisition

The simplest situation for single-channel recording is when the channel is active at a constant patch voltage, either spontaneously or as long as a stimulus is applied. In these instances data recording is not linked to voltage changes but to the experimenter's discretion or a stimulus such as a neurotransmitter, respectively. The immediate puzzle that this can create is that of data acquisition (see also Section 3.4.4). In contrast to voltage protocol responses, recording of (rare) spontaneous or stimulus-induced activity might require long recordings. It is not always correct simply to reduce the sampling rate because at the single-channel level the transitions might still be clustered in short periods of time. However, a fast sampling rate to capture these events, combined with long duration recordings, results in unwieldy data files. In general, recording of rare, fast activity is the most difficult. One way to reduce the data is to record only the openings, which involves the application of detector-driven recording (Figure 6.13). This requires quality recordings and a good data acquisition system. Otherwise, a compromise is inevitable. In most cases recording of non-voltage-dependent channel activity will start and stop by the experimenter's action, although some automation can take place depending on the experiment, e.g. if a trigger signal is sent to the computer by a fast perfusion device when it is activated. More often the responses are slow

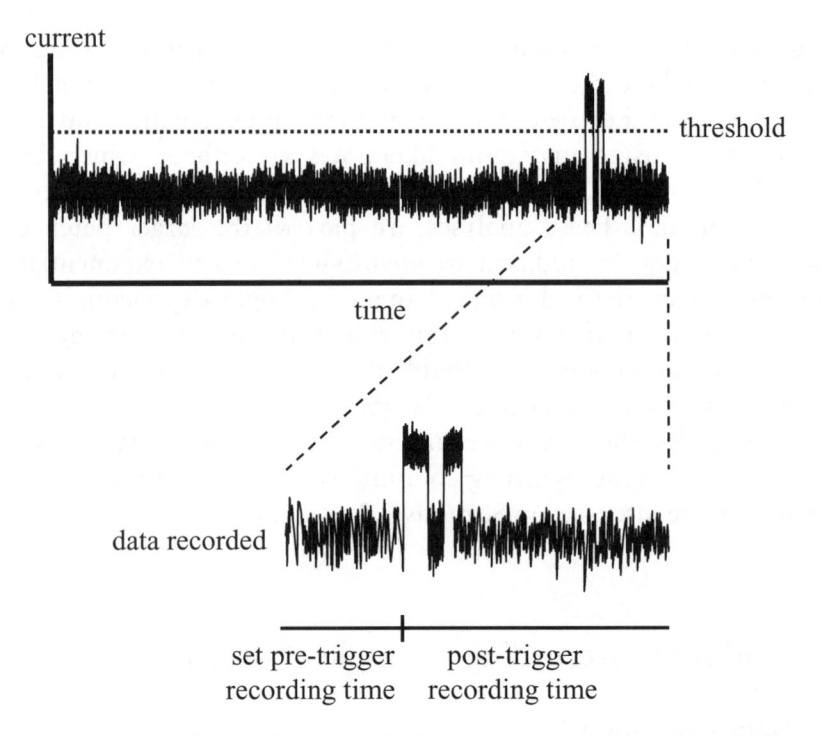

Figure 6.13 Under stable conditions single-channel data can be recorded selectively using a threshold and settings for pre- and post-trigger recording time to capture open dwell times

enough in onset (or spontaneous) to avoid complicating the set-up in this way.

6.2.2 Spontaneous activity

Spontaneous gating is an intrinsic property of ion channels owing to their stochastic behaviour. Spontaneous activity in the context of a single-channel recording could be defined as gating that is not induced by any particular stimulus, and is therefore recorded using continuous data acquisition. In excised patches such activity is likely to be truly intrinsic, because the experimenter can exclude any chemical stimulus from the intra- and extracellular milieu and check for voltage dependence. In the cell-attached patch configuration, 'spontaneous' activity might be driven by intracellular processes. These can include membrane potential changes

6.2 CONTINUOUS RECORDING OF SINGLE CHANNELS

because the membrane potential is not controlled in cell-attached patches. The way to trace the cause of the activity is to try to observe the same activity in excised patches, or to modify intracellular factors that you suspect of inducing the activity.

One of the most interesting forms of spontaneous activity in cell-attached patches is periodic activity, which is important in many cell types and physiological systems. The activity could be caused by oscillations in intracellular factors such as free calcium ions and reflect on membrane potential, or vice versa. Membrane potential changes show up in cell-attached patches by gradual changes in channel current amplitude at a constant holding potential owing to a changing driving force. (If you understand this sentence you have a good understanding of the cell-attached patch configuration, if not, please re-read Section 2.3.5 before moving on.) An example of such activity is shown in Figure 6.14. The size of the potential changes can be derived from the current amplitude changes if you have an idea of the slope conductance of the channel under study. This can be found through an I/V plot as discussed in Section 6.1.2. For example, an amplitude change of 1 pA through a 50 pS channel must be due (according to Ohm's law) to a membrane voltage change of 20 mV. To address the important question of whether or not the channel gating might *initiate* the observed membrane potential change, it must first be established that the gating is oscillation dependent, i.e. *P(open)* must be different for peaks in the oscillation compared to troughs. The difference must remain when the holding potential is varied, so it is unlikely to be a simple voltage dependency. Secondly, the current direction and open probability must combine to be consistent with a change in membrane potential in the observed direction at the observed times.

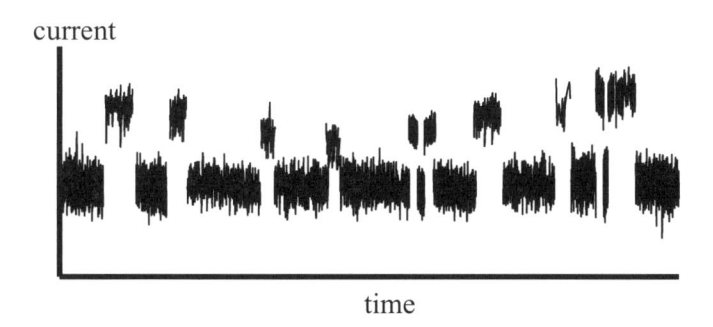

Figure 6.14 Oscillation in current amplitude in cell-attached patches is due to patch potential and subsequent driving force changes

6.2.3 Receptor-induced activity

Single-channel activity induced by receptor activation can be slow to develop, particularly when recording in the cell-attached patch configuration, and a multi-step signalling cascade is involved. As mentioned earlier, it is important in transient conditions to ensure that variables derived from the observed activity are stable over time or at least correlated with the time frame of the response. If responses are transient and relatively fast, then a ramp voltage protocol is an efficient way to obtain some basic characteristics of the channel under study. Voltage ramps for single channels are discussed in Section 6.3.2.

6.3 Study of Single-voltage-dependent Channels

Steady-state activity and current amplitude of voltage-dependent channels can be studied using continuous recording at different potentials, as discussed in Section 6.1. Voltage-dependent activation and inactivation, however, cannot be studied in this way and if the channel has rapid time-dependent inactivation, such as a TTX-sensitive sodium channel, you would not even see it under steady-state conditions. Hence, single-channel currents generated by voltage-dependent channels are, like voltage-dependent macro-currents, best studied using voltage manipulation protocols. There is, however, an important difference: In most cases macro-currents are the sum of a very large number of single-channel currents, creating a good picture of average activity at the set potential with only one trace. In single-channel studies, the behaviour of one protein is observed which – given the stochastic nature of the activity – can be very erratic. This is why in single-channel recording the voltage protocols must be repeated many times to create a reasonable average pattern of activity.

6.3.1 Step protocols

Single-channel step protocols are very similar to whole-cell step protocols but with the following differences:

- The voltage protocol must be adjusted to the configuration. The actual patch potential E_{patch} is simply the pipette (command) potential $E_{pipette}$ in outside-out excised patches but is $-1 E_{pipette}$ in inside-out excised

6.3 STUDY OF SINGLE-VOLTAGE-DEPENDENT CHANNELS

patches and $E_m - E_{pipette}$ in cell-attached patches (see also Sections 2.3.5, 2.3.7 and 6.1.1).

- Steps to each potential are repeated many times and if there are several channels of the same type in the patch with a reasonable $P(open)$, e.g. >0.1, then 50–100 steps might suffice to get a good average picture of channel activity during the step. If the channels open rarely at the step potential, more repeats are needed.

- Whole-cell-style transient correction (P/N subtraction, Section 5.2.2) is not always feasible in single-channel recording because of much higher signal-to-noise ratios and/or patch stability limitations to applied voltage steps. Instead, there is a P/N variation that sometimes can be applied that is particular to single-channel recording. If in a series of step repeats a number of steps do not evoke an opening, then those current traces will just consist of the passive response, i.e. capacitive transient and patch resistance. The traces can be averaged and subtracted from traces that do contain openings. The results (from a stable recording) will contain flat baselines (Figure 6.15). If there are many 'empty' traces, then the averaged correction trace will be very low in noise and provide better correction.

With respect to analysis, single-channel traces of current responses to voltage steps contain information for at least two additional descriptors compared with continuous recordings. The time it takes for the channel to open after a voltage step is known as latency, and could be considered to represent the *voltage-dependent* transition to the open state. Conversely, the time for the channel to close after the step can be seen as a measure of the voltage-dependent closure transition. These transitions can be kinetically different from the others that take place during the voltage step, although they might represent the same conformational states of the protein (Figure 6.16). A possible third descriptor is that of inactivation. The quasi-permanent closure of a channel during a voltage step might indicate time-dependent inactivation (linked to the time the channel has been open during the voltage step preceding this transition) or voltage-dependent inactivation (linked to the time the patch has been kept at that potential). Correlation analysis of the dwell times in relation to each other and the starts and stops of the voltage steps should distinguish between the two.

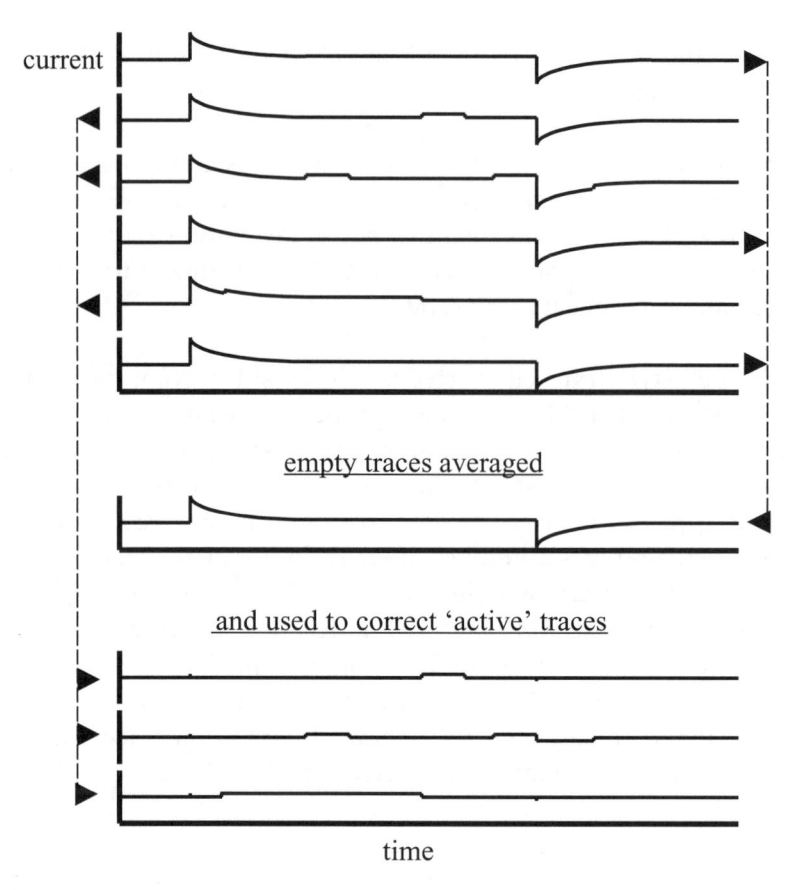

Figure 6.15 Positive/negative (P/N) subtraction for single-channel voltage steps. In a series of current response traces to a voltage step of constant amplitude, the traces without channel openings are averaged and used to eliminate the passive components of the traces with openings. Note the reversal in driving force after the end of the step in the second active trace

6.3.2 Ramp protocols

Ramp-shaped voltage changes can provide information about single-channel conductance and reversal potential very quickly. As with voltage steps and single-channels, ramps can be repeated many times to obtain an average voltage/activity relation, although a single trace with several transitions might be enough for an indication of the current/voltage relationship (Figure 6.17). In continuous recording this is particularly useful for transiently active channels, such as those activated by desensitising receptor mechanisms. Problems with membrane-constant-related de-

6.3 STUDY OF SINGLE-VOLTAGE-DEPENDENT CHANNELS

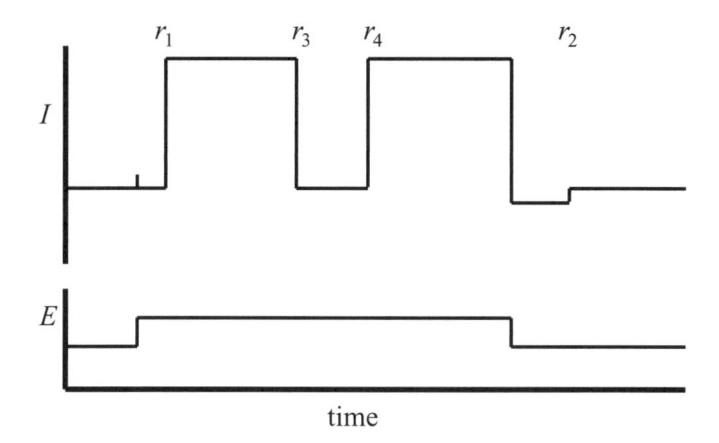

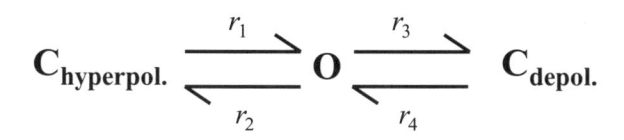

$$C_{hyperpol.} \underset{r_2}{\overset{r_1}{\rightleftarrows}} O \underset{r_4}{\overset{r_3}{\rightleftarrows}} C_{depol.}$$

Figure 6.16 The first transitions after voltage steps can be considered different from transitions within the voltage pulse. Although the underlying conformations can be identical (in this case, the closed state at negative potentials $C_{hyperpol.}$ can be the same as in the depolarised state $C_{depol.}$), the transitions are governed by different rate constants in voltage-dependent channels

lays, as encountered with whole-cell recording (Section 5.3.2), are practically absent so considerations regarding protocol design are usually confined to ramp speed in relation to voltage-dependent kinetic characteristics of the channel and to ramp size. Exact voltage-dependent gating descriptors such as activation and inactivation parameters cannot be deduced from ramps, but some idea of voltage dependency often can be obtained. The ramp speed has to be low compared with the voltage-dependent activation (but quick compared with inactivation, if present). Voltage-dependent gating will show as regions in the ramp responses where individual openings are more frequent than elsewhere, or rectification is visible in the averaged ramp response (Figure 6.18).

Another handy application of ramp protocols at the single-channel level is the quick initial scan after establishment of the desired patch clamp configuration. As for the amplitude histogram (Section 6.1.2), a ramp repeated ten times or so will show tidy levels like those in Figure 6.18 if there are one or more channels of the same type in the patch. If the patch

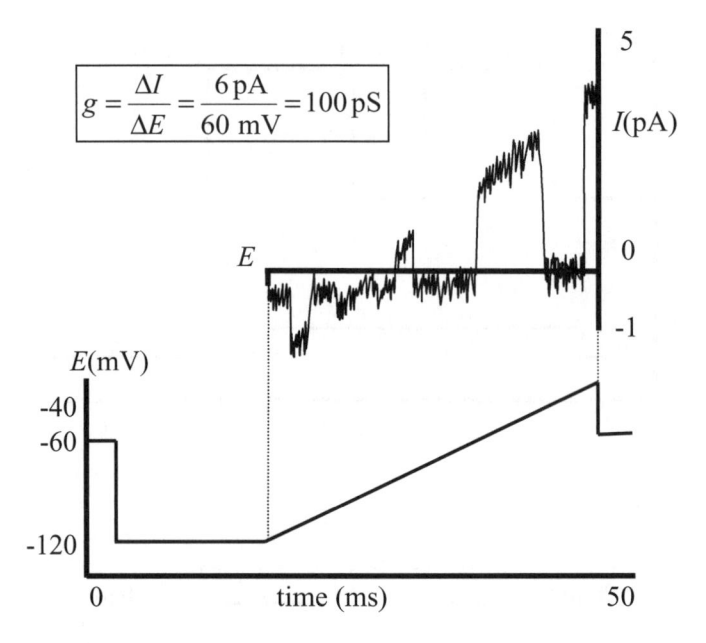

$$g = \frac{\Delta I}{\Delta E} = \frac{6\,\text{pA}}{60\,\text{mV}} = 100\,\text{pS}$$

Figure 6.17 A single ramp can give an indication of channel size and reversal potential. In this example of an outside-out patch with normal solutions, it is likely that the 100 pS channel is selectively permeable to potassium ions, because the reversal potential is very negative

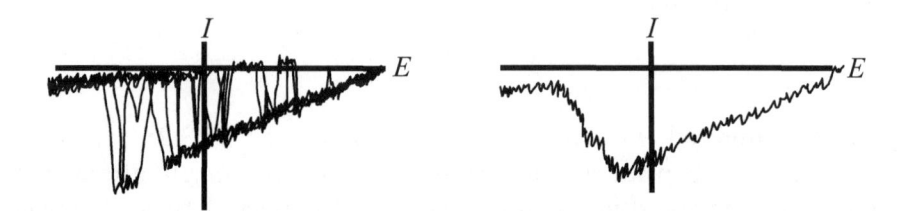

Figure 6.18 Voltage dependency of gating can be indicated by ramps. Superimposed traces (left-hand panel) show gating in certain areas of the ramp, reflected by non-linearities in the averaged trace (right-hand panel)

contains several channels of different amplitude, the picture will be chaotic.

6.3.3 Correlation with macro-currents

An often-followed path in the study of a population of ion channels is characterisation of the whole-cell current through those channels, in terms of voltage dependency of current amplitude, modulatory factors and ion specificity. Single-channel recording is then applied to obtain data that are difficult or impossible to acquire in whole-cell mode, such as any type of separation of gating and amplitude characteristics, detailed gating behaviour and data from studies under strictly controlled environmental conditions on either side of the membrane. The final step should be to correlate the single-channel data with whole-cell recordings. In practice, this is done by the old inductive method; the more characteristics that whole-cell and single-channel data have in common, the more likely it is that the single-channels mediate the whole-cell current. It is hard to say exactly what and how many characteristics the two data types should have in common, moreover because the absence of *conflicting* data is also a critical factor.

The comparison would be strongest when the recording conditions are equal. This can never be achieved because the whole cell allows local modulation (despite cytoplasm washout) that can only be matched by cell-attached patches, but then the ionic conditions are likely to be different. The safest parameter to compare is ion specificity, because this is usually a function of the channel pore only. The best situation for comparison is when conventional whole-cell recording is correlated with outside-out excised patch data, so that the ionic environments can be matched exactly. Things become uncertain when comparing voltage-dependent gating. An average of many single-channel responses to steps of constant amplitude should match the shape of a whole-cell current response using the same voltage pulse. If it does not, then there is still no guarantee that the channel does not mediate the whole-cell current:

- The whole-cell current could be carried by more than one channel type, including the one under study.

- The channel is modulated by an intracellular factor in whole-cell mode that is absent in excised patches.

If the profiles match, however, then it strongly suggests that the channel mediates the whole-cell current. This opens up the possibility for having some fun by feeding back the single-channel characteristics to the whole cell. For example, the combination of $P(open)$, the single-channel amplitude and the whole-cell current amplitude at a certain potential now allows calculation of the number of channels in the cell. Thus a whole-cell current of 800 pA, carried by a channel type that conducts 4 pA and has an open probability of 0.35, must be mediated by 570–580 channels.

Finally, any modulation by intracellular factors or otherwise should be similar for whole-cell and single-channel data. The most elegant situation in terms of progress is when the whole-cell and single-channel experiments inform each other in an ongoing learning process.

Further Reading

Textbooks

Aidley, D. J. and Stansfield, P. R. (1996) *Ion Channels – Molecules in Action*. Cambridge University Press, Cambridge, UK.
Brown, K. T. and Flaming, D. G. (1995) *Advanced Micropipette Techniques for Cell Physiology*. John Wiley & Sons, Chichester, UK.
Hille, B. (2001) *Ionic channels of Excitable Membranes* (3rd edn). Sinauer Press, Sunderland, MA.
Martin, R. (ed.) (1997) *Neuroscience Methods – a Guide for Advanced Students*. Harwood Academic Publishers, Amsterdam.
Ogden, D. (ed.) (1994) *Microelectrode Techniques – the Plymouth Workshop Handbook*. The Company of Biologists, Cambridge.
Rayne, R. C. (ed.) (1997) *Neurotransmitter Methods*. Humana Press, Totowa, NJ.
Sakmann, B. and Neher, E. (eds.) (1995) *Single-channel Recording* (2nd Edn). Plenum Press, New York.
Sherman-Gold, R. (ed.) (1993) *The Axon Guide for Electrophysiology and Biophysics Laboratory Techniques*. Axon Instruments, Foster City, CA.

Key Primary Literature

Barry, P. H. (1994) JPCalc, a software package for calculating liquid junction potential corrections in patch-clamp, intracellular, epithelial and bilayer measurements and for correcting junction potential measurements. *J. Neurosci. Methods* **51**: 107–116.
Becker, J. D., Honerkamp, J., Hirsch, J., Fröbe, U., Schlatter, E. and Greger, R. (1994) Analysing ion channels with hidden Markov models. *Pflügers Arch.* **426**: 328–332.
Blanton, M. G., Lo Turco, J. J. and Kriegstein, A. R. (1989) Whole cell recording from neurons in slices of reptilian and mammalian cerebral cortex. *J. Neurosci. Methods* **30**: 203–210.
Chung, S.-H. and Pulford, G. (1993) Fluctuation analysis of patch-clamp or whole-

cell recordings containing many single channels. *J. Neurosci. Methods* **50**: 369–384.

Church, J. (1993) A change from HCO_3^-–CO_2^- to HEPES-buffered medium modifies membrane properties of rat CA1 pyramidal neurones *in vitro. J. Physiol.* **455**: 51–71.

Cole, K. S. (1949) Dynamic electrical characteristics of the squid axon membrane. *Arch. Sci. Physiol.* **3**: 253–258.

Colquhoun, D. and Hawkes, A. G. (1995) The principles of the stochastic interpretation of ion-channel mechanisms. In: *Single-channel Recording* (2nd edn), Sakmann, B. and Neher, E. (eds.), pp. 397–482. Plenum Press, New York.

Donnelly, D. F. (1994) A novel method for rapid measurement of membrane resistance, capacitance, and access resistance. *Biophys. J.* **66**: 873–877.

Draber, S. and Schultze, R. (1994) Correction for missed events based on a realistic model of a detector. *Biophys. J.* **66**: 191–201.

Edwards, F. A., Konnerth, A., Sakmann, B. and Takahashi, T. (1989) A thin slice preparation for patch clamp recordings from neurones of the mammalian central nervous system. *Pflügers Arch.* **414**: 600–612.

Graham, J. and Gerard, R. W. (1946) Membrane potentials and excitation of impaled single muscle fibers. *J. Cell. Comp. Physiol.* **28**: 99–117.

Hamill, O. P., Marty, A., Neher, E., Sakmann, B. and Sigworth, F. J. (1981) Improved patch-clamp techniques for high resolution current recording from cells and cell-free membrane patches. *Pflügers Arch.* **391**: 85–100.

Hille, B. (1972) The permeability of the sodium channel to metal cations in myelinated nerve. *J. Gen. Phys.* **58**: 599–619.

Hodgkin, A. L. and Huxley, A. F. (1952) Currents carried by sodium and potassium through the membrane of the giant axon of *Loligo. J. Physiol.* **116**: 449–472.

Kangrgra, I. M. and Loewy, A. D. (1994) Whole-cell patch-clamp recordings from visualised bulbospinal neurons in the brainstem slices. *Brain Res.* **641**: 181–190.

Ling, G. and Gerard, R. W. (1949) The normal membrane potential of frog sartorious fibers. *J. Cell. Comp. Physiol.* **34**: 383–396.

Meech, R. W. and Standen, N. B. (1975) Potassium activation in *Helix aspersa* neurones under voltage clamp: a component mediated by calcium influx. *J. Physiol.* **249**: 211–239.

Neher, E. and Sakmann, B. (1976) Single-channel currents recorded from membrane of denervated frog muscle fibres. *Nature* **260**: 799–802.

Radden, E., Behrens, M., Pehlemann, F. W. and Schmidtmayer, J. (1994) A novel method for recording whole-cell and single channel currents from differentiating cerebellar granule cells *in situ. Exp. Physiol.* **79**: 495–504.

Ravesloot, J. H., Van Putten, M. J. A. M., Jalink, K. and Ypey, D. L. (1994) Analysis of decaying unitary currents in on-cell patches of cells with a high membrane resistance. *Am. J. Physiol.* **266**: C853–C869.

Rohlicek, V. and Schmid, A. (1994) Dual-frequency method for synchronous measurement of cell capacitance, membrane conductance, and access resistance on single cells. *Pflügers Arch.* **428**: 30–38.

Sakmann, B., Edwards, F., Konnerth, A. and Takahashi, T. (1989) Patch clamp techniques used for studying synaptic transmission in slices of mammalian brain. *Quart. J. Exp. Physiol.* **74**: 1107–1118.

Spruston, N. and Johnston, D. (1992) Perforated patch-clamp analysis of the passive membrane properties of three classes of hippocampal neurons. *J. Neurophys.* **67**: 508–529.

Spruston, N., Jaffe, D. B., Williams, S. H. and Johnston, D. (1993) Voltage- and space-clamp errors associated with the measurement of electrotonically remote synaptic events. *J. Neurophys.* **70**: 781–801.

Surmeier, D., Wilson, C. J. and Eberwine, J. (1994) Patch-clamp techniques for studying potassium currents in mammalian brain neurons. *Methods Neurosci.* **19**: 39–67.

Ueno, S., Ishibashi, H. and Akaike, N. (1992) Perforated-patch method reveals extracellular ATP-induced K^+ conductance in dissociated rat nucleus solitarii neurons. *Brain Res.* **597**: 176–179.

Velumian, A. A., Zhang, L. and Carlen, P. L. (1993) A simple method for internal perfusion of mammalian central nervous system neurones in brain slices with multiple solution changes. *J. Neurosci. Methods* **48**: 131–139.

Wilson, W. A. and Goldner, M. M. (1975) Voltage clamping with a single microelectrode. *J. Neurobiol.* **6**: 411–422.

Yawo, H. and Chuhma, N. (1993) An improved method for perforated patch recordings using nystatin–fluorescein mixture. *Jpn. J. Physiol.* **43**: 267–273.

Index

access resistance, 39–40
action potential, 17
activation curve, 132
activation, 132
AD/DA converter, 90
air tables, 47
aliasing, 88
alligator clip, *see* crocodile clip, 79
ampere, 13
amphoteracin B, 35
amplitude histograms, 144
amplitude resolution, 86, 91
analogue filter, 89
analogue-to-digital converter, 89
anti-vibration tables, 45
antifungals, 108
axon, 17

baclofen, 138
bath electrode, 64
bathing solution, 33
beeswax, 86
Bessel, 88
bilayer, 5
bioelectricity, 1
bleach, 66
borosilicate glass, 69
brain slices, 2, 51
Butterworth, 88

CA1 neurones, 51
$Ca2^+/Na^+$ exchanger, 9
cable properties, 17, 124
cadmium ions, 126

caesium ions, 126
calcium fluorescence, 51
capacitance, 13
    compensation, 25, 81
capacitive transient cancellation, *see*
        capacitance compensation, 82
capacitive transients, 99
capacitor, 15
carbonate buffer, 96
cell membrane, *see* plasma membrane, 5
cell wall, 6
cell-attached patch, 34
cell-attached patch mode, 32
channel modelling, 154
charge, 13
chloridation, 67
clipping, 92
clock pulses, 77
closed state, 143
co-transporter or exchanger, 9
command voltage, 30
concentration gradients, 7
conductance, 115, 13, 1
continuous recording, 137
coulomb, 13–14
crocodile clip, 79
current, 13–14
    clamp, 28
    injection, 27
current–voltage relationships, 127
cytoplasm, 5, 8
cytosol, *see* cytoplasm, 23

dampening characteristics, 44

dampening properties, 46
DAT recorders, 137
data acquisition, 90
decibels, 88
dielectric constant, 16
differential amplifier, 24
diffusion, 7
digital filter, 89
digital noise, 77
digital-to-analogue conversion, 92
digitisation, 90
discriminator, *see* event detector, 90
drift, 43
drift correction, 152
driving force, 13-14
dual-electrode voltage clamp, 30
dummy load, 79
dwell-time analysis, 144
dwell time distributions, 155

earth, 14
electrical field stimulation, 52
electrical interference, 76
electrical parameters and units, 13
electrode glass, 69
electrogenic, 9
electrophoresis, 62
endoplasmic reticulum, 9
energy barriers, 13
entropy, 7
equilibrium potential, 8
equivalent circuits, 18
event detection, 148
event detector, 90
excised patch configurations, 41
external noise, 76, 84
extracellular electrodes, 18
extracellular solution, 96

f noise, 85
f^2 noise, 85
$1/f$ noise, 85
farad, 13
Faraday cage, 51
filtering, 81
filter, 87
 characteristic, 87

band-pass, 87
cut-off frequency, 87
high-pass, 87
low-pass, 87
notch, 87
orders, 88
poles, 88
slope, 87
fire polishing, 71
flat noise, 85
flood management, 50

G protein, 12
$GABA_B$ receptor, 138
gating, 11
Gaussian distributions, 144
Gaussian filter, 88
gentle kiss, 106
(Gibbs) energy, 8
gigaseal, 95
glutamate receptors, 132
ground, 25
 electrode, 33
 loop, 80

HEPES, 96
hippocampal slice, 51
holding potential, 29
Hum, 76
hydraulic manipulator, 52

I/V relation, *see* current-voltage
 relationships, 127
idealised traces, 153
immunofluorescence, 51
impedance conversion, 89
in series, 21
inactivated state, 135
inactivation, 129, 132
 curve, 135
infrared light, 48
inside-out excised patch, 34
interference contrast microscopy, 48
intracellular
 and extracellular medium, 5
 recording, 1
 solution, 96

intrinsic noise, 80
inverted microscopes, 49
inward rectifying potassium channels, 29
ion, 9, 2, 5
    channel, 9, 2, 5
    selectivity filter, 12
    specificity, 144

Kirchoff's current law, 20
Kirchoff's voltage law, 19
Köhler light source, 48

lamp noise, 50
latency, 161
leak current, 15
leak resistance, 38
ligand-gated channels, 11
liquid junction potentials, 74
liquid–liquid junction potentials, 65
local circuit currents, 17
local superfusion, 52
low-pass filters, 83

macro-current recording, 34
macro-currents, 160, 40
Markov behaviour, 155
mechanical manipulator, 52
membrane
    capacitance, 27
    potential, 1
    resistance, 40
microelectrode, 64, 2
micromanipulator, 47
micropipette perfusion, 62
micropipette puller, *see* pipette puller, 68
micropipettes, 23
microscope, 43
    platform, 50

Na^+/K^+ pump, 9
Na^+/K^+-ATPase, 9
Nernst equation, 8
nicotinic acetylcholine receptors, 132
NMDA channels, 29
noise analysis, 86
noise prevention, 84
noise reduction, 76

Nomarski microscopy, 48
non-selective cation channels, 132
normal distributions, *see* Gaussian
    distribution, 145
nystatin, 35

ohm, 13
Ohm's law, 15
oocytes, 2
open probability, 152
open state, 135
osmolarity, 7, 96
osmosis, 6
outside-out excised patch, 34
outward rectifier, 129

parallel, 21
passive components, 117
patch clamp amplifiers, 81
patch resistance, 41
Peltier element, 61
perforated patch clamp, 31, 2
perforated patch, 34
    recording, 108
periodic activity, 159
peristaltic pumps, 61
pH, 96
phase shift, 88
phase-contrast microscopy, 48
phosphorylation, 12
phospholipids, 5
pipette
    capacitance, 27
    electrode, 64
    holder, 67
    pressure, 56
    pullers, 70
    resistance, 23
    solution, 24
plasma membrane, 5
platform, 43
platinum, 66
polarisation of electrodes, 65
P/N subtraction, *see* positive/negative
    subtraction, 121
positive/negative subtraction, 119
potential, 13

power, 76
  grid, 76
  spectrum, 85, 87
  supply, 50
pressure system, 47
probe, 24
puffer system, 62
pulse protocols, 116
pump, 9

quartz glass, 69

radiation noise, 77
rate constant, 156
RC circuit, 22
receptors, 5
reciprocal rule, 20
rectification, 148
redox potentials, 65
relative permeabilities, 131
resealing, 113
resistance, 13
resting current, 116
rms noise, 85
ROCs, *see* ligand-gated channels, 12

salt bridge, 58
sampling rate, 90
search mode, 103
second messenger operated channels, 12
secondary active transport, 9
selective permeability, 12
series resistance, 40
  compensation, 82
sharp electrode, 29
shielding, 77
short-circuiting, 21
siemens, 13
signal conditioning, 119, 81, 84
signal-to-noise ratio, 84
silicone gel, 86
silicone tubing, 61
silver chloride, 66
single-channel
  configuration, 32
  recording, 34
single-electrode voltage clamp, 31

slope conductance, 128
SMOCs, *see* second messenger operated channels, 12
sodium-linked glucose transporter, 10
sojourns, 152
solid–liquid junction potentials, 65
space clamp, 28, 30
squid giant axon, 1
stability, 43
steady-state inactivation, 135
step protocol, 127
stochastic behaviour, 158
sub-conductance states, 147
superfusion, 1
switch noise, 77
sylgarding, 86
synaptic potentials, 1, 30

tail current protocol, 128
temporal resolution, 90
tetraethylammonium, 126
tetrodotoxin, 125
time constant, 22
time-dependent inactivation, 120, 160
torque, 55
transporter, 9
Tschebycheff, 88
TTX, *see* tetrodotoxin, 126

unblock, 134
universal gas constant, 8

vibration, 43, 53
video camera, 51
VOCs, *see* voltage-dependent channels, 12
volt, 13
voltage
  clamp, 1, 15, 28
  ramp, 130
  sensors, 11
voltage-dependent channels, 11

Walther Hermann Nernst, 7
washout, 38
water jacket, 59
white noise, 85

whole-cell, 34
    configuration, 29
    mode, 33

window current, 136

zap control, 83